Transportation Issues, Policies and R&D

Transportation Infrastructure

Assessment, Management and Challenges

TRANSPORTATION ISSUES, POLICIES AND R&D

Additional books and e-books in this series can be found on Nova's website under the Series tab.

TRANSPORTATION ISSUES, POLICIES AND R&D

TRANSPORTATION INFRASTRUCTURE

ASSESSMENT, MANAGEMENT AND CHALLENGES

SAÚL ANTONIO OBREGÓN BIOSCA
EDITOR

Copyright © 2018 by Nova Science Publishers, Inc.

All rights reserved. No part of this book may be reproduced, stored in a retrieval system or transmitted in any form or by any means: electronic, electrostatic, magnetic, tape, mechanical photocopying, recording or otherwise without the written permission of the Publisher.

We have partnered with Copyright Clearance Center to make it easy for you to obtain permissions to reuse content from this publication. Simply navigate to this publication's page on Nova's website and locate the "Get Permission" button below the title description. This button is linked directly to the title's permission page on copyright.com. Alternatively, you can visit copyright.com and search by title, ISBN, or ISSN.

For further questions about using the service on copyright.com, please contact:
Copyright Clearance Center
Phone: +1-(978) 750-8400   Fax: +1-(978) 750-4470   E-mail: info@copyright.com.

NOTICE TO THE READER

The Publisher has taken reasonable care in the preparation of this book, but makes no expressed or implied warranty of any kind and assumes no responsibility for any errors or omissions. No liability is assumed for incidental or consequential damages in connection with or arising out of information contained in this book. The Publisher shall not be liable for any special, consequential, or exemplary damages resulting, in whole or in part, from the readers' use of, or reliance upon, this material. Any parts of this book based on government reports are so indicated and copyright is claimed for those parts to the extent applicable to compilations of such works.

Independent verification should be sought for any data, advice or recommendations contained in this book. In addition, no responsibility is assumed by the publisher for any injury and/or damage to persons or property arising from any methods, products, instructions, ideas or otherwise contained in this publication.

This publication is designed to provide accurate and authoritative information with regard to the subject matter covered herein. It is sold with the clear understanding that the Publisher is not engaged in rendering legal or any other professional services. If legal or any other expert assistance is required, the services of a competent person should be sought. FROM A DECLARATION OF PARTICIPANTS JOINTLY ADOPTED BY A COMMITTEE OF THE AMERICAN BAR ASSOCIATION AND A COMMITTEE OF PUBLISHERS.

Additional color graphics may be available in the e-book version of this book.

Library of Congress Cataloging-in-Publication Data

ISBN: 978-1-53614-059-0

Published by Nova Science Publishers, Inc. † New York

Contents

Preface		vii
Chapter 1	Transport: Accessibility, Urban Development and Mobility *Saúl Antonio Obregón Biosca*	1
Chapter 2	The Experience of Active Mobility and Its Contributions to Urban Habitability *Avatar Flores-Gutiérrez, Guillermo I. López-Domínguez, and Verónica Leyva-Picazo*	45
Chapter 3	Advanced Assessment and Strategic Management of Paved Road Networks: The Case of Costa Rica's National Road Network *José D. Rodríguez-Morera and Luis G. Loría-Salazar*	71
Chapter 4	Airport Pavement Management *Mauricio Centeno*	103

Chapter 5	Effective Assessment and Management of Railway Infrastructure for Competitiveness and Sustainability *José A. Romero Navarrete, Frank Otremba and Saúl Antonio Obregón Biosca*	**121**
Chapter 6	Urban Policies for Sustainable Mobility: The Return of the Tram to Barcelona *Pere Macias, Antonio Gonzalez, Abel Ortego, Elisabet Roca, Robert Vergés, Josep Mercadé, Joan Moreno, Alessandro Scarnato, Ana María Moreno, Jesús Arcos, Clement Guibert, Marianna Faver, Etienne Lhomet and Míriam Villares*	**149**
About the Editor		**183**
Index		**185**

PREFACE

Transport infrastructure are considered a key factor in the economic and social development in the territory of any country. Their cause–effect relationship makes development possible, although they do not provoke it directly (because cannot replace the traditional means of production, like work and capital). Although transport infrastructure do not cause the socioeconomic growth directly, they bring about the possibility of changes in the patterns of population distribution and directly support activities of production.

The transport infrastructure generates benefits in efficiency and interchange of goods and people. The increase of accessibility induced by transport infrastructure in a region causes employment growth, which causes a more positive economic future for the affected area by the infrastructure. At the same time, the change in the accessibility will influence the resident population, because it modifies mobility and induces travel demand on the road network.

On one hand, there is an attempt to meet the need for housing, educational and health services, and public transportation systems as well, but on the other hand, this attempt generates problems such as chaotic and poorly planned cities with uncontrolled growth. Some researchers have worked to demonstrate an important relationship between transport systems and infrastructure, socioeconomic growth, industrial location and land use, that is to say from an urban point of view, the improvement in accessibility has direct repercussions on the standard of living of the people involved.

The purpose of this book is shown to the practitioners, researchers and students a comprehensive framework around the transport infrastructure for the following:

- Transport infrastructure and their economic influence and in the territorial transformation.
- The active mobility infrastructure, the design and the habitat influence in urban areas.
- The fundamentals and main approaches around the road infrastructure, pavement management.
- The fundamentals and main approaches around the road infrastructure, pavement design, assessment and management in roads and airports pavements.
- An approach around the assessment and management of railway infrastructure and the urban policies in tram systems in Europe.

The transport infrastructure diversity presented in this book offers a valuable and representative point of view about their importance, considering from the assessment aspects, management and especially the challenges in the field, in the most representative transport infrastructures.

The list of contributions is not exhaustive; and hope that our colleagues, professionals, researches, as well as students, find this book a useful contribution around the transport infrastructure world. Additionally, the material is complemented by a selected overview of advanced case studies and applications.

In: Transportation Infrastructure
Editor: S. Antonio Obregón Biosca

ISBN: 978-1-53614-059-0
© 2018 Nova Science Publishers, Inc.

Chapter 1

TRANSPORT: ACCESSIBILITY, URBAN DEVELOPMENT AND MOBILITY

Saúl Antonio Obregón Biosca[*]
Universidad Autónoma de Querétaro, Querétaro, México

ABSTRACT

Transport engineers are involved in the planning, design, construction and management of transport infrastructures, for the transport of goods and people. Transport infrastructure is a key factor on the economic development of a country or region, and influence the territorial transformation, so appropriate infrastructure planning will induce higher social profitability. The accessibility changes promote the activities location, allows the territorial transformation and the change on the land use values, the profitability of the public transport systems and how accessibility improving influences the mobility patterns and can induce the economic and social development. In this sense, the chapter theoretically addresses the accessibility concept promoted by the transport infrastructure and how it induces: i) the territorial transformation, ii) the socioeconomic development (including a section about the relationship of transport and new economic geography), iii) the

---

[*] Corresponding Author Email: saul.obregon@uaq.mx.

urban form and commutes, and iv) the transport impacts and approach of the transportation modeling. The main challenges and research approaches are defining, around a general point of view that introduces to the transport infrastructure context.

Keywords: accessibility, planning, transport infrastructure, socioeconomic effects, territorial transformation

INTRODUCTION

Transport infrastructure promote the economic and social development of a territory. Their cause–effect relationship makes development possible, although they do not provoke it directly (the infrastructure cannot replace the traditional means of production, like work and capital). Although transport infrastructure do not cause the socioeconomic growth directly, they bring about the possibility of changes in the patterns of population distribution and directly support activities of production (Obregón-Biosca, 2010; Obregón and Junyent, 2011).

The economic growth (higher per capita income and numbers of workers) induces demand for new housing (Bhatta, 2009), and these new developments require new transportation infrastructure (Bhatta, 2010). The urban development is "the process of emergence of the world dominated by cities and by urban values" (Clark, 1982). In this sense, the urban growth is the "spatial and demographic process"; and the urbanisation "is a spatial and social process which refers to the changes of behavior and social relationships that occur in social dimensions as a result of people living in towns and cities" (Bhatta, 2010). The travel is an activity that takes place from one given geographic location to another, over transportation network (Oppenheim, 1995). The accessibility term is used to mean a variety of measures of how well connected are at given location with activities of a given type of: work opportunities, shopping destinations, residential, commercial areas and many others, usually in terms of how long and how much of a given aimed activity is located (Badoe and Miller, 2000). Accessibility is an aspect key of location.

Physical accessibility is determined by time and travel cost to other locations. It depends on presence, efficiency and effectiveness of transport modes. Investment in new transport infrastructure will alter relative accessibility of locations inducing socioeconomic changes. Thus, the property market acts as conduit through where economic and social impacts of accessibility changes are transmitted to build environment (Badoe and Miller, 2000).

In this context, the chapter address the accessibility concept induced by infrastructure and transport systems, as well as some indexes. This accessibility induced by infrastructure and transport systems influences the trips and in the urban form, reorganizing mobility patterns, at the same time influencing the activities location. Finally, the chapter addresses the impacts induced by infrastructure and transport systems and introduce in their modeling.

THE TRANSPORT SYSTEM AND THE ACCESSIBILITY

The transport system must comply the main needs of citizens in the sense of the possibility of each citizen to mobilize and access at their needs in a city or region (Pardo, 2005). Ozbay et al. (2006) states that a reliable transportation system and efficient infrastructure are necessary for the population economic well-being, as it provides access to the region and efficiency allows of commerce, industry, labor and housing.

Accessibility has been analyzed from diverse perspectives, both its definition and measurement tools, focusing on the opportunity that people has at a certain point to participate in a particular activity or set of activities (Thakuriah, 2001). Deficiency accessibility can be enable discrimination, in sense of marginalization and decrease of well-being or life quality. Another approach considers the separation (temporal or spatial), so accessibility is focused as measure of the access facility. Under this approach Harris (2001) exposes symmetric access: "If A has access to B, then B has access to A," but also mentions that its measurement can be asymmetric in space, defining the inaccessibility, "the opposite of easy access." Fernández (2000) expose it, according to the use and enjoyment approach of all people, and not only by a segment. Different concepts of accessibility have emerged recently, for example, the accessible environment (obstacles removal in physical environment) and the universal design (universal provision by all people), and both described in Buhalis et al. (2005). In the sense of the territorial planning, the accessibility rate is an opportunity based on distance, cost or time. In this sense Cerda and Marmolejo (2010) argues that accessibility is not only determined by transport networks but mainly mobility is physical travel when the people move through the city. Henry (1998) conceptualizes mobility as the way in which the people performs activities program in their spatial and temporal dimensions. Thus, mobility materializes at spatial level in the mandatory trips by the daily needs of individuals linked to the ease of network access and transport systems.

Transport strategies are often evaluated using accessibility rates which researchers and policy makers can easily manage and interpret, however, more complex and disaggregated accessibility rates can difficult the interpretation (Geurs and Van Bee, 2004). For example, a new public transport service could be provide a significant improve of the accessibility; contrarily, if society who are not disposed to use public transport, and/or if public transport use is restricted to people with low income (Martinez, 2008). Ryan (1999) emphasize the need to consider who are the users, and who benefit when accessibility improve.

The Accessibility Index

Transport systems impact the life quality; and the accessibility is a major point in the planning and assessment (Obregón et al. 2016a). Accessibility has been analyzed from diverse perspectives both its definition and measurement terms, mainly on the opportunity that a person have to participate in an activity or activities (Thakuriah, 2001). The indices depends of their input data, and lead to diverse results, so the index choice depends of the situation and its purpose (Guy, 1983 and Handy and

Niemeier, 1997). In this sense, it is necessary to consider the aim and the parameters that include each one with the purpose of choose the most consistent and accurate for the study object. Guy (1983) exposes that the results may markedly different depend the index, so in Handy and Neimeier (1997) argued that the situation and the study aim defines the index.

Regard the accessibility indexes in Garrocho and Campos (2006) exposes that can be classifieds in five categories: i) spatial separation, is the distance average of the commutes between origins and destinations and the demand sensitivity in transport cost changes (distance friction factor); ii) spatial interaction, estimates the service units attraction considering the time, distance, cost, among others, as spatial parameters in a friction factor between zones. That is to say, zone attraction is considered as a factor; iii) cumulative opportunities, considers the destinations as service units, and the travel time or access to potential destinations as an accessibility measure; iv) utility, based on the individual perception of each user to each unit service as destination; and v) spatial-temporal, based in the temporal restrictions of the individuals (as potential users). The individuals have limited time to make their activities, that is to say, for a long travel time less time to make their activities. Considers the limitations of the individual's performance (time to rest, sleep or eat, as well ages differences). Too considers, the "authority" restrictions as legal or normative mandates that inhibit movement or activity. For example, on one hand, the spatial separation index considers the distance as the variable between the origin-destination and the demand sensitivity to the changes in transport costs (friction factor). It estimates the average of the paths between all origins to all the possible destinations.

$$A_i = \frac{\Sigma_j c_{ij}}{d}$$

where: A_{ij} accessibility index (spatial separation); c_{ij} the transport cost between O-D pairs, and d the friction factor. On the other hand, the spatial interaction index considers the supply dimension (as attraction factor) and the transport cost (susceptible to variations), and can be estimated using the following expression.

$$A_i = \sum_j \frac{O_j}{c_{ij}^d}$$

where: A_i accessibility index (spatial interaction); O_j the attractiveness of the unit service; c_{ij} the transport cost between O-D pairs, and d the friction factor.

The five accessibility categories refers to the population, service units or both, as discussed in Bach (1981), while in Garrocho and Campos (2006) defined as components: i) the physical, considers an a parameter of the geographical distance between the supply and the demand (service units - users), and ii) social, the parameter of the social distance between the service and the user (revealed accessibility to the services), and need surveys to collect the preferences (Allun and Phillips, 1984). Nevertheless, Geurs and Ritsema van Eck (2001) define four specific components: i) transport, it's the utility of an individual to cover a distance between a origin and a destination, considering the travel time, costs and comfort level; ii) land use, explains the activities geographical distribution; iii) temporal, focused in the opportunities along the day and the available time to do the activities; and iv) individual, focuses in the needs considering the socioeconomic characteristics of the individuals, their opportunities and their abilities, to reflect how these parameters influence the access level to transportation and the skills to qualify in jobs close to their residence.

In addition to the components and categories exposed, have several approaches, in Arentze et al. (1994) and Handy and Niemeier (1997) three are exposed: i) the travel cost, measures the access to arrive in a transport mean (considering the commute cost) to the activities in the diverse land uses; ii) gravity, approach based on the trip behavior making an analogy to the physical model of gravity, where the interaction of available masses for the consumers, be or not elected; and iii) restrained based, sustain that the individual has a spatial and temporal dimension, and commute limitations are not only affected by the distance but also by the available time of each person.

In the last sense, Miller and Wu (2000) argued that space-time delimits the destinations that an individual can reach, considering the time duration of the mandatory activities and the travel time of the transport system, basis on the Lenntorp (1976) approach, shown that the evaluation of a routes set to the destination (service), considering the activity schedules and spatial restrictions. From this discussion Miller and Wu (2000) propose two space-temporal approaches in the accessibility index: i) surplus utility, incorporates the individual characteristics of each traveler who choice the one that maximizes its utility, thus considers the random utility theory proposed by McFadden (1974); and ii) the composed, combines the time-space component and the utility models, that is to say, the utility to do discretionary trips considering mandatory mobility and time (available and of the commute). However, accessibility index must accomplish criteria that determine their consistency, in this sense Weibull (1976) and Morris et al. (1979) state that: i) the origin-destination data order does not affect the index; ii) index response should not be increased since transport costs or reduced by offer increase; iii) it is necessary to consider the individuals spatial patterns; iv) operationally feasible (not an academic example); and v) clear result interpretation. Bath et al. (2000) argue that the presented criteria are fundamental references for the accessibility index design.

From the exposed, different parameters are required depending the index, from the available information to the technology and technical capabilities, while the aim defines the most appropriate for the study case. Therefore, Handy and Clifton (2001) recommended to consider: it) important parameters of the population, ii) data availability; iii) planner's technical ability to make sense the information, and iv) zoning, allows detailed analysis of the surroundings qualitative characteristics. The most commonly applied indexes are spatial separation and spatial interaction.

THE URBAN FORM AND COMMUTES

The urban spatial structure induces the origin-destination separation, while commutes are the result of home and activities location (Valero, 1984). In this sense, Gutiérrez and García (2006) shows that the territorial transformations in metropolitan spaces and their regions, induce to reach more sprawled destinations and far from the home. From the above, the connection between the urban spatial structure and the transport is established, which is indicated in the time-duration and trip distance. In this sense Obregón et al. (2015) expose that the metropolization phenomenon induces the sprawled human settlements. The immersed areas in this territory shown functional connectivity to its central core. Paaswell and Zupan (2007) establishes that in such places a high accessibility is required that is provided by the transport systems. Urban sprawl is a serious problem in the sense that induce major resources demand (drink water, electricity, sewage, paving, among others), and transport services. Two urban models has been argued, the monocentric and the polycentric, that define the internal structure of the activities in a metropolitan area, and both consider as a determining factor the transport.

Fuentes (2009) recognizes in the monocentric model the transport system importance within the city internal structure definition, stating that the Central Business District (CBD) is an export node that concentrates the major employment places of the city, and at the same time induces two problems: the first, the workers travel time to CBD, so their location is extremely important. Second, the land price, the proximity to CBD results lower transport cost, and in the opposite sense in the periphery. The polycentric model is defined by García (2010) as "the process by which a society gradually moves away from a spatial structure characterized by the existence of a single employment center, moving towards a new spatial structure where several employment centers coexist in the same or different hierarchical order." However, is not clear in the polycentric model, the number and centers size (Guillermo, 2004), and its effect on urban mobility (Song, 1992). In this sense, Camagni (2005) establish the city term as *"machine informationnelle,"* that is to say, a machine that builds and reproduces itself "manufacturing its own programming," "a significant machine that gathers and connects, the productive chains, institutional and scientific." It seems that the best way to represent a city is through the environment connectivity.

One point that has been study object in the metropolization phenomenon is the mandatory mobility, and as Sobrino (2007) exposes the theory and empirical studies indicate that "home location is strongly determined by the workplace, but is unclear the causation sense." Likewise, transport system improvements and accessibility changes induced by infrastructure are elements that intervene in the behavior of commuting. Therefore, Malayath and Verma (2013) argue that policies should focus on the people's mobility behavior and change the way than they travel through a different environment. On the inhabitants side, Azócar et al. (2010) argue that as a city grows, not only the urban weave becomes complex and fragmented, but also the sociocultural composition of its residents.

Miller and Wu (2000) exposes that the commute behavior focus in the individual understanding of processes and decisions, from the factors analysis that determine the activity of each trip. Consequently, in transport planning are considered, how or when travel activities will do along the transport system, emphasizing accessibility with alternative approaches to travel behavior in the transport system performance. In this sense, the accessibility provision to evaluate the individuals participation in activities, it is possible to focus on the opportunities and travel cost. Cerda and Marmolejo (2010) argue that one of the variables that impulses a discontinuous urban system is accessibility, which is commonly measured in terms of cost, time or distance. The above is related to the transport network, therefore, they consider necessary a mobility-based approach, as this is the physical reflex of people's mobility through the city.

Rodríguez (2012) report the polycentrism progress and the spatial sprawl of employment in the Metropolitan Area of Santiago (Chile), observing a mixed process in this area, in the sense of the development of new subcenters that not counter the central core predominance. Considering the relationship between city size and employment in six French regions, Schmitta and Henry (2000) analyzed the correlation between the size of the city, the influence of the rural population and changes in employment. Their results shown that the central core size and their employment growth rates have influence on the rural population and its workplace change, mainly in medium-sized urban centers. On a commercial location approach, Escolano and Ortiz (2005) analyze how these activities disrupted the monocentrism in the city of Santiago (Chile). They considers the distance to the CBD, the population geographical distribution and their income, accessibility and road network. Their results shown the change process that has broken Santiago's traditional monocentric model, and could respond to a mix of the mono-centric and the polycentric model (mono-polycentric), in which the center maintained some control, but the subcenters extends its influence, to another smaller ones. Gallo et al. (2010) analyze the Madrid region restructuring considering that the patterns of sprawl and polycentrism are simultaneous. The authors conclude that there is a mixed model, that is, the subcenters consolidation connected by high accessibility induces dynamics in those influence areas that structure the region. In the same sense, Pengjun (2013) analyzes the peri-urbanization trends after the year 2000 in Beijing (China). He observed that the temporary migration (mainly, young people with high studies level) in the area increased, as well as the social inequality. The author warns the vertical and horizontal fragmentation, so it will not be easy to achieve the social integration between urban-rural, concluding that it is necessary to promote planning capacity thus minimizing segregation.

To analyze urban and regional transport systems, the accessibility is essential, due to the increasing complexity of transportation systems and their impact in the individuals life quality, for this reason, it is common to use accessibility variables in transportation planning and assessment. In this sense, Alonso (2007) argues that accessibility involves city effective planning, requiring useful and accurate data to estimate future trend scenarios. Its definition must be associate to both the region zones in study and the service units, possibility measuring, therefore, according to Goodall (1987) definition, "accessibility is the ease with which can reach a place (destination), from other territory points (origins), so it synthesizes contact opportunities and interaction between origins and destinations," in this sense Johnston (2009), accessibility is synthesized as "the interaction and contact opportunity between origins and destinations," according to Thakuriah (2001), focuses the accessibility concept in the opportunity for a individual in a particular place to participate in a particular activity or activities set. From this definitions, two aspects stand out, first the possibility to reach some service between origins and destinations. Second, the interaction between individuals and destinations.

Couch and Karecha (2006) expose about urban sprawl debate among European planners is often expressed in terms of urban contention and the compact cities search. This debate also sustained in Latin America and mainly research reported focuses on the evolution urbanization patterns, and how transport is linked. For example, Azócar et al. (2010) identify and analyze the patterns of Coyhaique city (Chile), which is characterized by a geographic fragmentation and isolation, among other aspects. Their results shown that the result of fast urbanization is the location advantages which are linked to the geographical position of the cities respect the transportation infrastructure. Couch and Karecha (2006) analyze the British policies to urban sprawl control in the Liverpool conurbation. Interested in how market forces can be redirect towards the compact cities, based on the Brehny (1997) studies, they doubt about the viability and the acceptability of a compact zone when one aspirations of population, is home-locate in the suburban living. Analyzing secondary information and household surveys conducted with this arguments, they conclude that although recent developments in British policies, market forces continue been the major challenge for policy makers to promote the compact city.

Farhad (1996) exposes that the coordination lack in transport and land use planning promoted an expansion model of several metropolitan areas in the United States of America. Therefore, analyzes why and how these variables should be reoriented to comply American laws on the efficiency of intermodal transport (ISTEA) and clean air (CAAA). Their results provide parameters for those responsible for public policies involved in transport planning and land use. Proposes four conditions for a new metropolitan vision: i) have the necessary authority, as well as the financial capacity and technical and experience to plan, ii) have the support and cooperation of public administrations, citizens and private promoters, iii) overcome the fragmented governance problem that occur in many metropolitan areas, and iv) local governments should review and coordinate their land use plans, zoning regulations, plans and programs to reach the aims of land use plans and mobility. Grengs (2010) considers the hypothesis that spatial mismatch start from the difference between modes of transport, analyzes the Detroit Metropolitan region by a gravitational model. Proves the access differences to jobs and provides support to reconceptualize spatial mismatch. They show that while Detroit presents the greatest distances of workers to US jobs, central neighborhoods reflect an advantage of access to jobs relative to the rest of region, they conclude that the flexibility to accommodate the distinctive features of a region related with their level of development can be an effective means of access improving.

Alonso et al. (2013) analyze how population and employment growth influences public transport planning in Madrid (Spain), focus on the investment in high capacity systems such as light rail (Metro) specifically in the network that connects the capital with the neighbours municipalities. Their results show that the low investment cost in the light rail reduced the financial problems in the Madrid government and the new transport services demand increased during the first years, although it later decrease. Argue that their results suggest that the complexity of economic decisions and policies are linked to growth, adapting to diverse development policies around the city. Quayle and Driessen van der Lieck (1997), McGranahan et al. (2004) and Pengjun (2013) consider that there is no agreement on the solutions or approaches needed to promote the peri-urbanization sustainable development, since this process is complex and depend in each country.

The Urban Sprawl and Accessibility

The process of gradual spread out of urbanization has given way to the term "urban sprawl" (Bruekner and Fansler, 1983), involving a transport situation characterized by passing through vacant land from one area to another (Hamidi and Ewing, 2014; Zhang, 2004; Bhatta, 2010). Undesirable effects of such a phenomenon include the sacrifice of farmland (Brueckner and Largey, 2008) and the reduction of social capital (Putman, 2000), whereas the administrations have to solve the associated infrastructure-related problems concerning sanitation, drinking water, and transport.

Bhatta (2010) states that urban sprawl is a concept with difficulties in its definition, but that a general consensus is that the sprawl "is characterized by unplanned and uneven pattern of growth, driven by multitude of processes and leading to inefficient resource utilization." In this same sense, Rahman et al. (2008) explains that this area is characterized by an adverse urban development that interferes with the urban environment and is neither an acceptable urban situation nor suitable for an agricultural-rural environment. Duany et al. (2001) suggest that there is urban sprawl because of the invention of the automobile and that it increased after World War II. Following this way of thinking, Bhatta (2010) affirms that most of the cities have experienced or are experiencing the sprawl phenomenon, including cities in developing countries. Bengston et al. (2005) discussed the negative effects of the sprawl, arguing that urban sprawl has aroused a wide social focus because it can avoid regional sustainable development. Sudhira and Ramachandra (2007) include the change in land use and the cover of the land as negative effects because sprawl induces building and paved areas. Klug and Yoshitsugu (2012) remark that the urbanization is always accompanied by the development of physical infrastructure, which requires huge investments and determines the structure of a city over long periods of time, which cannot be easily adjusted to changing patterns of the services demanded. The satisfaction of the induced mobility needs and consequent traffic would utterly affect the environment (Bhatta, 2010).

Rural environments show problems related to accessibility, transportation and mobility to service centers. Nutley (2003) highlighted the absence of public transport in rural environments, leading the participation in vehicles. The foregoing, induce interurban disparities, mainly in travel times to work, the use of car or public transport, Fuentes (2009) argued that the differences are product of the differentiated accessibility to the work centers. Through a spatial analysis they determine that the accessibility, the land value of and the number of jobs in secondary sector the variables that explain the increments in the travel times. In his research he consider on the activities location, because in the central cores higher rent are paid than in periphery, however, the worker pays a high transport cost, which is corroborated in Obregón et al. (2016b) where it estimates the travel generalized cost, emissions and fuel consumption induced by the urban sprawl. In the same sense, Currie et al. (2009) reported that car ownership in households is higher in peripheral areas, concluding that influence the social exclusion and the car dependence.

The work and school commute involving the nuclei and the peripheral areas has become a social and economic problem to analyse because it has been reported that the residential location is strongly influenced by the workplace location (Sobrino, 2007). However, it should be noted that a cause-effect relationship is unclear, i.e., whether the residents first got the job and then choose their place of residence or vice versa. Nevertheless, according Sanchez (1999), as referenced by Houston (2012), lower skilled workers tend to commute shorter distances, which introduces another parameter that should be assessed in future research.

In this context, Garcia (2010) suggests that raising mobility in urban sprawl should take into account not only the number of trips but also the travel times and distances involved, recognizing that large-distance trips are less frequent than short-distance trips. The dimensional changes of the sprawling metropolitan areas originate territorial and social adaptations because a growing number of out of the nucleus destinations are engaged for a wider range of activities.

That is, the development of transport and communications has lengthened the daily distances travelled, multiplying the mobility choices for a large part of the population in all aspects of urban life: the place of residence, location of activities and work, and for socializing purposes (Ascher 2004 cited in Garcia 2010).

The Metropolitan Mobility

Hiernaux (2005) states the mobility of contemporary societies confronts us to evaluate the identity concept in its spatial dimension, as the central variable to reconstruct the modern individual and the territorial dynamics. On metropolitan expansion and mobility, Lizarraga (2012) analyzes the Metropolitan Area of Caracas (Venezuela), which shown a high residential segregation as regards income and inequitable mobility conditions and accessibility, stating in part as result of deregulation and privatization of public transport. Exposes that the Metro or Metrocable projects have not helped to reduce negative externalities and the social exclusion, therefore analyze the mobility characteristics, concluding in proposals to promote the sustainable urban mobility, for example: the extension and improvement of safe transport infrastructures for pedestrians and cyclists, as one way to increase their accessibility; the decrease of vehicle use and the public transport reorganization. In Mexico, Suárez and Delgado (2010) report the residential mobility in the Mexico City observing evidence of a process of co-localization between jobs and housing. By a statistical analysis they predict the probability of residential changes when there the workplace changes. Their results show that exists influence of the work place on the household choice, and for this reason they define that the mobility makes a social equilibrium mechanism of the urban structure. Ortiz and Escolano (2013) analyze how the evolution from a compact city to a diffuse city induces functional changes, and how contributes to modify the social segregation scale. Analizing the city of Santiago (Chile), they observe how the higher socioeconomic groups are concentrated in administrative entities of the city, and were not valued like permanent residence places by these sector, then this phenomenon has led to a major complexity of the general model of city residential segregation.

Gakenheimer (1998) reports an overview of the relationship between mobility, motorization rate and demographic increase, analyzing several mega-cities in developed countries, which show high population growth and urban sprawl, and conclude that the dysfunctionality are between urban design and motorization rate.

Soto and Álvarez (2012) analyze the process between conurbations and metropolitan areas in the Gran Valparaíso region (Chile), shown that this tendency induces asymmetries among the different territorial units from the perspective of the employers of workplaces and household areas, conclude that these territorial phenomenon explain the differences in the work commutes. Each sub-area within a zone shown dissimilar mobility characteristics, as reported in Millward and Spinney (2011), who analyze the travel time differences in the urban-rural continuum. Four urban categories delimited in Halifax County (Nova Scotia). Its main result shown that the residents near the urban area shown higher travel times. They obtain an average travel rate of 6.8 (one-way), in time average are 96.6 minutes, exposes that are higher than been reported in surveys, like in Murakami and Wagner (1999) and Stopher et al. (2007), the high value as result of the survey characteristics, which is called STAR (Space Time Activity Research, which was the first large-scale application of a GPS-assisted survey). As expected, the time consumption tends to move from the center to the city outside, and the trip production is less, in response to fewer opportunities and higher the distances. They exposes that the population compensate the upper travel times in the suburbs by reducing their commutes: the trip number decreases significantly in the suburbs. Trip averages in the suburban are similar, and the number of trips are significantly less in the outermost ring (rural zone), being a sprawled activity place.

On the mobility policies in the peri-urban with low income populations Jouffe and Lazo (2010) analyze the case of Santiago (Chile), explaining that the public transport system is inadequate for these inhabitants and induces a socially marginalized territories. Hiernaux (2005) assumes that the mobility increase does not identities destroy but rather reconfigures them through new territorial dynamics.

TRANSPORT AND THE NEW ECONOMIC GEOGRAPHY

The new economic geography (NEG) indicate that the trade benefits can approach in a convergent or divergent concentration, which responds of the regional characteristics and explained by transport costs, agglomeration forces and the economic activity dispersion (Sánchez-Reaza, 2010). The NEG bases its theory on potential market studies, the cumulative causality and central places theory. In Fujita and Thisse (1996) the spatial equilibrium is conceptualized as a scenario in which no company is stimulated to change its location because it would imply looking for a different location to obtain greater benefits, then the agglomeration and dispersion forces (centripetal and centrifugal) provide the guidelines to locate as long as a homogeneous general equilibrium is induced.

Fujita et al. (1999) expose that in the centripetal forces economies of scale, transport costs, effects on market size, market access and products stand out; all linked to a central or strategic location. Centrifugal forces are related to internal diseconomies such as congestion and pollution, the land rent cost, the inputs dispersion, competition for prices and the labor cost. Mendoza-Cota and Pérez-Cruz (2007) explain each one of the two forces. Since a transport infrastructure improvement serves as the link to integrated the market, allowing better connections on the areas of lower economic activity, it is established that the transport infrastructure improvement induce an attraction for the industrial location (Holl 2004a; 2004b), while Alañón and Arauzo (2008) concluded that reducing the access time induces positive effects in the location. The model used by Holl (2004a) takes into account quality factors and improvements in road infrastructure, in addition to location, cost, demand, and agglomeration economies.

$$\pi_j = \sum_{k \in P} \sum_{k \in M_r} p_{kj}(T_{jk}^D) q_{jk} - c_j(w_j, g_j(T_{jk}^S), q_i) - f_i$$

The firm expects profits from a representative municipality j, which will depend on the sum of the expected revenue from sales in all markets and local production costs. By selling the output produced in municipality j, q_{jk} in place k within region r, the firm obtains an average turnover of P_{jk} (T^D_{jk}), which depends on transport costs, due to location j, k, T^D_{jk}. Thus, revenues from sales in different places are not only determined by the size of the market for firms in these places, q_{jk}, but also by access to these markets T^D_{jk}. It assumes that since firms cannot incur transport costs in the sale of production within the municipality itself or income in the local market, it will merely depend on price and demand.

Ríos and Obregón (2017) in the study case of the Mexican Bajio Region, established that the accessibility induced by transport infrastructure in the regional market, to the national market and import-export ports does not prove to be the variable with greater significance in comparison to the other variables that approximate to explain the manufacturing location in that region. That is to say, the transport infrastructure is a necessary but not sufficient factor in the industrial location. Their results highlight the variables that describe the accessibility potential of to the demand (interregional and intraregional), exhibit a highly significant approximation. These variables explain the manufacturing location in the region together with characteristics, as: i) the local market size; ii) urbanization economies; iii) land rent cost, and iv) the schooling average degree.

TRANSPORT IMPACTS

The effect of the highroad is provided in stages as noted in Sutton (1999), who analyzed the long-term impact on land use caused by the peripheral ring of I-225 in Denver. He observed three stages in the evolution of land use. The first stage was significant residential development together with small business and offices. The second stage saw an increase of shops and offices, and the third stage saw the development of high density buildings of offices.

Ozmen et al. (2007) analyzed the factors affecting companies' relocation decisions in a study of the region that included 21 counties in

New Jersey representing the "destinations" of the new businesses, whereas New York and Philadelphia represented the "origins." They suggested a "gravity based" business relocation model, developed and calibrated using an iterative approach. Their results show that businesses moving from Philadelphia tend to locate deeper into New Jersey than businesses moving from the New York area. This result might be because businesses are attracted to the high density of the New York market and its economic and cultural potential, and therefore they do not want to relocate as much. The results of the sensitivity analysis and estimated market elasticities also support this finding. Hanson (1986) related the transport system and urban dynamics. She realized the importance of the choice of residential location and workplace, which are linked to three factors: the topology and the quality of housing, neighborhood characteristics, and the level of accessibility. From a social perspective, Robert Cervero (1999) examined the accessibility of urban areas through job and housing opportunities and the transport network within the area of San Francisco. The main conclusions of Cervero showed social inequality in the accessibility of the labor areas. He observed a growing separation of job and residential locations in the process of American suburbanization, principally in the lower social strata. Cervero (2003) discussed how the investment in highway construction stimulate the multiplication of trips. Using information from 24 highways in California, he notes that adding lanes to these highways provokes an increase in traffic because of the advantage in speed of travel. Coray and Manoj (2009) develop a general methodology to model the effects of socioeconomic factors in highway construction and expansion to examine impacts on low-income families.

Siccardi (1986) explores the economic effects of highway construction specifically as a result of economic development funds that have been available to the states from the mid-1970s to 1980s, and he discusses the activities of the states he believes to have utilized the funds most effectively on the basis of the real or perceived economic effects of the effort. Bérion (1998) notes that, in the sixties, it was thought that the construction of major transport networks would significantly help the regional development of new territories (structural effects). In the seventies, new investigations showed the permissive factors caused by the highways. The author concludes that the infrastructure makes the regional development possible but did not directly cause it. Other French authors focused their studies on the analysis of the relationship between infrastructure and regional development. Among them one can cite Plassard (1978), who studied the structural effects of highroads construction from a spatial perspective. Dubois-Taine (1991) studied the structural effects induced by the roads on the territory.

Burmeister and Joignaux (1997) authored a book of the major studies of territorial restructuring caused by the transport system. It can be concluded that the road network by itself does not affect the territorial transformation (in terms of economic development), but it changes the patterns of population distribution and directly supports productive activities.

One of the most complete works regarding economic impact is by Banister and Berechman (2000). They presented a book that addresses the effect on economic growth brought about by transport infrastructure; it contains a review of research in the areas of economic, social, spatial, and environmental changes. At the same time, it takes an analytical approach to the technique of modelling the effects of transport infrastructure at the macro level, in local development and economic growth, and the evaluation of projects. At this point, the writers conclude that the vast majority of research has focused on the macroeconomic impact. Although it is possible to establish statistical relationships at this level, it is difficult to build relationships that support causal links, such as the effects of external factors (time and stage of development) that may influence the direction and force of impact. Finally, the writers present empirical case studies about the economic impact of highways, rail, and airports. One of the most relevant conclusions is the indisputable evidence of the changes in accessibility on the economic development, changes in demographics, the location of activities, and employment, among others.

Employment growth implemented by transport investments is a result of the interaction of two main factors. The first is the impact on the willingness of a worker entering the job area for travel to work once travel costs have been lowered. The second relates to the demand from employers and the level of access to a skilled workforce. Berechman and Paaswell (2001) analyzed the effects of the change in accessibility to work. The idea that improving accessibility in an area (a result of an investment in transport infrastructure) will better enhance participation in productive activities depend on factors such as socioeconomic characteristics and work area location. They examined the effects of improving the accessibility by transportation investments on the supply of labor in the South Bronx (New York), an area that is economically depressed. This area has been undergoing an improved transportation system, known as the "Center of the Bronx." The research showed the positive impact of reduced transport costs on the labor market. They concluded that the impact will be deeper in areas of lower income, since, for this population, the cost of transportation is a real barrier for the worker. Recently, Ruiz and Kockelman (2006) applies a random-utility-based multiregional input-output model to assess the Trans-Texas Corridor impacts on trade, production, and worker locations. The model predicts a slight redistribution of economic activities, increasing the supremacy of counties located closer to export zones. It also suggests a greater diversification of economic activity/production and moderate changes in the distribution of wages, floor space rents, and population following the production trends. Ozbay et al. (2003) investigate the impact of accessibility changes on the level of economic development in a given region.

Correa (2010) states that the public transport system works in a systematic way, as everything that moves in a city, and therefore, any intervention in any transport mode will affect the rest of the system. This author exposes the public Transantiago transport system case (in Santiago, Chile) and explains that after its opening, the mobility policy focuses on making investments in isolated projects. Therefore, it criticizes in the sense of investments rationalization and evaluates them, stating that urban transport projects must stop being the administrations emblematic projects, and therefore, evaluate and decide from a systemic and organized view on the territory, urban development and mobility, searching the optimum social. Gutiérrez (2000) analyzes the Buenos Aires Metropolitan Region

(Argentina), taking in to account the structure and dynamics of the metropolitan region, with the aim of establishing a relationship between transport policy and social conditions in the territorial structures changes. It recognizes structural changes, this being guided by commercial agents and resulted from the State passive attitude, which engagement the public service in the future.

Malayath and Verma (2013) argue that the economy boom in emerging countries promote the well-being and the way of life, and inevitably the private vehicles dependence, together with population growth, the increase in vehicle ownership promotes to the travel demand with its negative consequences (accidents, congestion, pollution, inequality, among others). Newman et al. (1995) expose that one of the policies to reduce automobile dependence is physical planning. They discuss the planning policies problems, pricing, the moral dimension and hope for the cities. Gathering these policies, they conclude on the importance of the physical planning. Considering Santiago city (Chile) Ureta (2009) analyzes the motive and perception behind the acquisition and car use, in order to define public policies that encourage sustainable forms of urban mobility. Proposes some elements that must be considered, as: i) understanding that ownership and use are different, ii) mitigate the automobile habits, especially in terms of commuting, iii) observing the car-use as part of a great sociotechnical system, iv) socializing the real costs of the car, both personal, social and environmental, v) overturning the "myths" of the car and vi) introduce a gender perspective.

In the sustainability sense Malayath and Verma (2013) exposes that the present urban transport scenario that leads to the development and promotion of sustainable transport policies. Considering the relationship between commutes and greenhouse gas emissions Miralles (2012) shown that the automobile contribute about 30 percent of total emissions, but not all means of transport have the same level, for that analysed the modal split in Catalonia (Spain) paying attention in the urban - rural and the travel motive. Modarres (2013) calculate the energy consumption in some communities of the Metropolitan Area of Los Angeles, under the assumption that political attempts to achieve greater density and greater balance between jobs and housing should consider the social geography of metropolitan areas and their close relationship with energy consumption patterns. Their results confirm the urban density importance in determining trip patterns and energy consumption, noting that it is important to pay attention to these areas within a metropolis.

Modelling Transport

Martínez (2008) recognized the less understood interaction between land use and transport, and how the and what extend transport projects induce the urban development. McNally (2008) exposes that the basic behavioral defining the interaction centered on the activities that the individuals wish to participate, but Martinez (2008) described that the activities are spatially dispersed describing a land use pattern, and to reach them, the individuals need to travel, for that, propose how understand the activities are located in the territory and the interaction between the land use and transport. To understand it, proposes on one hand, a model to describes the behavior the consumers, land developers and builders, and predicts the urban development as a result of their interaction considering the microeconomic paradigm of the consumer`s behavior under two urban economic approaches, the first, the classical utility maximizing assumption to describes consumers with rational behavior, and in the urban context the choices are discrete (the random utility approach has reported by McFadden, 1978). The second, the bi-auction approach proposed by Alonso (1964) and assumes that the urban land is assigned to the highest bidder, that is to say, the existence of location access advantages, which can be capitalized by owners in the form of higher rents selling their properties in auctions; thus the location is assigned to the highest bidder. From above mentioned, Martinez (1992) proposed include the bid-choice location that share the common assumption that consumers are utility maximizers but differ the formation of land rents. Considering Alonso (1964) willingness to pay function defined as the inverse in land rents of the correspondent indirect utility function V conditional on the location choice as:

$$WP_{hvi} = I_h - V_h^{-1}(P, Z_{vi}, U_h^0)$$

where: I_h is the income of the locator agent, P is the vector of goods prices, z_{vi} is the vector of attributes (including access), P the vector of good prices, U^0 the utility level. In this sense represents the maximum value the consumer is willing to pay for a location (v,i) described by z_{vi}, to obtain a utility level U^0 given a fixed I_h and a exogenous P. Jara-Diaz and Martínez (1999) consider in the function include the access attributes (acc) represents the consumer's need to interact with the activities, and include a variety of non transport users' externalities called traffic nuisance (η) and the location externalities, denoted by z_{vi} depend on the build and natural environment, that is to say, represent the interaction between consumers' choices is expressed by the mutual dependency between willingness to pay functions as:

$$WP_{hvi} = I_h - V_h^{-1}(P, z_{vi}(WP), U_h)$$

From above, the bid-choice approach in Martinez (2008) exposes that have the advantage of recognizing a very special characteristic of land, and is relevant in the urban context because the effect is seen in rent differentials across the cities, and the access measures is a key point to understand the land use and transport interaction. According to Willumsen (2008) the "travel demand is manifested in space and time and to model this one must be represent the supply of transport infrastructure and services in some formal way." In this sense, exposes that a transport network is an analytical construct that facilitates the identification of routes fooled by travelers and their corresponding costs, and be formally represented as set of links and nodes. To modelling of bus, rail or other forms of transport require the detailed identification of routes served, stops, access modes and times. The transport modelling provides the tools to describe and predict the movements of persons, goods and information in a given or possible future environment (Axhausen, 2008). Ortúzar and Willumsen (2008) exposes that it stablishes relations between the amounts, firms, infrastructures, services, locations, characteristics and behaviours of persons. According to Axhausen (2008) the focus of transport modelling is the movement of persons, goods and patterns and intensity of private and commercial telecommunication, these movement are collected by surveys or observation; the movement implies a definition of activity. An activity is a continuous interaction with the physical environment, a service or person, within the same socio-spatial environment.

Some countries development national models to make forecasts of traffic and travel demand. For example, the Netherlands development a national model to forecast traffic on the strategic road and rail networks. It was originally developed for the national road infrastructure planning, but the model soon evolved to transport policies like environmental and railway planning. In Italy, the Ministry of Transport standardized a methodology to support policy information to be applied for all modes of transport. In Thailand, a conventional four stage structure model to predict trips by private vehicles, train, bus and air. Daly and Sillaparcharn (2008) exposes that the national models be useful to the planning apparatus and the countries are continuing to invest and development of existing national models and further countries are development new ones.

The highway maintenance policy as defined by Rouse and Putterill (2008) as the actions directed towards preserving and enhancing the integrity, serviceability and safety of highways, aimed at offsetting ongoing physical surface and sub-surface changes. Is the third major cost driver, in flexible pavement setting, these would include pavement age, reseal cycle, adequacy of drainage, chip type and size. Some methods are employed, as i) life cycle cost management, that managing costs efficiently over the life of the asset; ii) the scale and efficiency effects from amalgamation, to focus to increase the mean average road network length, and iii) the environmental factors as cost drivers, like the sub-surface characteristics, rainfall, slope, among others. These methods emphasize the important interrelationship of cost and quality of services.

On the freight transport, some mathematical models was development focused on the strategic freight network planning. Friesz and Kwon (2008) exposes that they are applied primarily to forecast, months or years in the future the freight traffic over specific network links and routes and though specific network nodes and terminals, considering a multimodal partial equilibrium of the transport market, with alternatives being evaluated according to the comparative static paradigm. In this sense, the freight network models are not based on statistical inference, because they have the important capability of examining the implications of structural changes in underlying markets, that is difficult to do with econometric methods, but not impossible. The demand of freight transportation services is derived from the spatially separated production and consumption activities associated with individual commodities. Friesz et al. (1998) idealized freight network planning model in 15 criteria as: i) multiple modes compete for, and used to carry freight shipments; ii) freight transportation involves multiple commodities with distinct transportation cost characteristics and diverse time requirements; iii) sequential loading of commodities, it is sometimes possible to prioritize commodities and assign them individually to the network in order from highest to lowest shipment priority; iv) simultaneous loading of commodities, the disaggregation schemes will lead, however, to commodities of identical shipment priority but with distinct unit cost characteristics; v) the variation of relevant costs and delays with flows volumes due to congestion; vi) elastic transportation demand, the fact of demand will generally vary with transportation costs and delays; vii) explicit shippers and carriers, the route and modal choices in freight systems result of the decisions of shippers and carriers that obey distinct behavioral principles and sometimes, have goals conflicting; ix) sequential and simultaneous shipper and carrier submodels, whether one ascertains the decisions of the shippers first and then the decision of the carriers or determines both simultaneously; x) sequential and simultaneous computable general equilibrium (CGE) and network models, general equilibrium models employ assumptions about freight transportation costs, and the network model in an attempt to produce consistency, and only simultaneous solution result in the desired consistency; xi) non-monotonic functions, for cost and delays that are expected to occur of average rail operating costs which initially decline as volume increases and then begin to increase as capacity approached; xii) explicit backhauling, recognizes that a large portion of traffic is made up of empty rolling stock, barges and trucks, that contribute to costs and congestion; xiii) blocking strategies, rail freight flows are composed of trains of varying length, made up of different types of cars that frequently blocked in groups bound for common destinations; xiv) fleet constraints, the restrictions on the supply of rolling stock and vehicles that cannot be violated in the short run, and xv) imperfect competition, the tendency of carriers to collude with one another and to bargain with shippers in setting rates. Friesz and Kwon (2008) exposes that these criteria depend on the dichotomy of freight decision-making agents: shippers and carriers.

Nash and Smith (2008) states that has been a strong interest in measuring performance of rail operators, and some empirical studies concerned with the impact of privatization, de-regulation and vertical separation on the performance. In this sense, exposes that the performance measurement is difficult in the railway sector, but have a considerable interest to policymakers. The usual starting point for the measures of rail such as passenger kilometres and freight tonne-kilometres., but the outputs need to described in terms of the provision of transport of a specific quality from a specific origin and destination at specific point in time, that is to say, different products have significantly different cost characteristics. Some approaches of productivity measurement are employed, like the partial productivity measures, that compare the ratio of a single output to a single input across firms and over time. The total factor productivity measures, is a traditional approach that measure the ratio of all outputs to all inputs. The advantage of the index is the large number of distinct outputs and inputs maybe identified. The econometric approaches are based in Caves et al. (1980), and their application and comparison are reported in Caves et al. (1981) and Caves et al. (1982). In this sense, Nash and Smith (2008) exposes that the lack of a statistical test concerning the impact of the institutional environment on performance, for that son efficiency-based approaches to performance measurement are reported in

Oum et al. (1999) and Smith (2006) provided detailed coverage of the various efficiency studies for the sector.

In the sense of airport performance two aspects have attention, the congestion processes and costs, and the merits of different options such as pricing, administrative control and investments to reduce it, from above, Forsyth (2008) exposes that "recent models of airport delays and pricing have opened up what had seemed to be a fairly settled issue." Considers that the central problem is the data, and with different inputs international comparisons are hard. For that reason, need it to estimate more robust models to determine the efficiency and relate it with the operational and environmental factors that maybe influence the productivity.

Shipping industry are categorized in two major sectors, the bulk shipping (raw materials) and the liner shipping (final and semi-final products). The models developments in this industry according to Haralambides (2008) are focused on two types: i) the optimization of liner shipping operations, and ii) in liner shipping as a market structure, the economic modeling of market structures and tariff setting processes. In this sense, the phenomenon of global shipping alliances thus emerged to exploit economies of scope among otherwise competing operators, through strategies.

DISCUSSION

It is not possible in an overview of this type to cover all the relevant aspects around the transport, urban development and the accessibility concept. But as discussed, one of the major concerns in making decisions on investment in transportation infrastructure has been to ensure that there are socioeconomic benefits and promoting the sustainable territorial development. Although the travel times and fuel consumption are clear, understanding the economic effect presents difficulty in some areas of theoretical and empirical cases. Transport infrastructure promotes development and will depend on the conditions in each region and the possibility of generating investment projects. Common conclusions from the authors discussed here are that: roads are one of the biggest providers of social and economic welfare.

On the territory, the transport infrastructure restructures its possible accessibility to new developments, in this sense the transport infrastructure can induce territorial effects like the sprawl, the externalities derived from urban sprawl could be used as a practical reference for public agencies and urban planning professionals when designing and implementing measures aiming at mitigating the negative effects. In this sense, exist a clear correspondence between the need of transport infrastructure and the urban development. The literature in effects of infrastructure, in land value shown that is determining the transport infrastructure effects inducing socioeconomic growth improved accessibility and land uses. Transport accessibility improve it and are socioeconomic impacts such as increased land values due to better location, proximity to service centers, reducing travel time, travel costs, fuel savings, wealth redistribution, poverty reduction, redistribution of population, its respective activities and improving access to employment centers. In that sense potentiates spatiotemporal relates the choice of destinations and everyday behavior (i.e., changes in mobility patterns). The re-structuring and territorial structure transport infrastructure network is the main element of spatial, favoring aperture space and decentralization.

The inhabitants and the activities in the territory, are involved to determine the need of infrastructure, in this sense, the transport modelling is an important part of most decision-making processes, and focuses to simplify and abstract the important relationships underlying the provision and use of transport. Models are focuses, for each kind of infrastructure but the common denominator to the approaches adopted is the deployment of a mixture of a synthesis of methods. In this sense, need to know the likely implications of their actions on the transport network.

On the sense of industrial location, the literature shown in this chapter reflects the importance of accessibility. Although the firm's industry suggests that geographical specialization has also been a major force, the road network and land availability also play a key role in industry location. The importance of the transport infrastructure in economic activity influences the localization of companies in productivity increases and markets reorganization promotes tourism while facilitating access to points of interest and generates new employment. Therefore, transport infrastructure influence distribution of income and redistribution of activities in the region benefitting the growth of economic activity. And also help to reduce possible economic regional disparities and has a direct effect on the quality of life for inhabitants inside the territory served by transport infrastructure.

REFERENCES

Alañón, Á. and Arauzo, J. M. (2008). Accesibilidad y localización industrial. Una aplicación a las regiones españolas fronterizas con Francia. *Revista de estudios regionales,* 82, pp. 71-103. [Accessibility and industrial location. An application to the Spanish regions bordering France. *Magazine of regional studies*]

Allun E. J. and Philips, D. (1984). *Accessibility and utilisation: Geographical Perspectives on Health Care Delivery*, London, Harper & Row, 214 p.

Alonso, F. (2007). Algo más que suprimir barreras: conceptos y argumentos para una accesibilidad universal. *Trans, Revista de traductología*, 2(11), pp. 15-30. [Something more than removing barriers: concepts and arguments for universal accessibility. *Trans, Journal of Translation Studies*]

Alonso-Neira, M. Á., Gallego-Losada, R. and Pires-Jiménez, L. (2013). La ampliación del Metro en la periferia de Madrid (1999-2011). *EURE*, 39(118), pp. 123-148. [The expansion of the Metro in the periphery of Madrid (1999-2011)]

Alonso, W. (1964). *Location and Land Use*. Cambridge, Harvard University Press.

Arentze, T. A., Borgers A. W. and Timmermans, H. J. P. (1994). Multistop-based measurements of accessibility in a GTS environment.

International Journal of Geographical Information Systems, 8(4), pp. 343-356.

Ascher, F. (2004). *The new principles of urbanism*, Alianza Ensayo, Madrid, Spain.

Auxhausen, K. W. (2008). Definition of movement and activity for transport modelling. In: D. Henser and K. Button (eds.) *Handbook of Transport Modelling* (2nd edition) Elsevier, The Netherlands, pp. 329-343.

Azócar-García, G., Aguayo-Arias, M., Henríquez-Ruíz, C., Vega-Montero, C. and Sanhueza-Contreras, R. (2010). Patrones de crecimiento urbano en la Patagonia chilena: el caso de la ciudad de Coyhaique. *Revista de Geografía Norte Grande*, 46, pp. 85-104. [Patterns of urban growth in Chilean Patagonia: the case of the city of Coyhaique. *Geography Magazine Norte Grande*]

Bach, L. (1981). The Problem of Aggregation and Distance for Analyses of Accessibility and Access Opportunity in Location-Allocation Models. *Environment and Planning A*, 13, pp. 955-978.

Badoe, D., and Miller, E. (2000). Transportation-land-use interaction: Empirical findings in North America, and their implications for modeling. *Transportation Research Part D*, 5(4), pp. 235–263.

Banister, D. and Berechman, J. (2000). *Transport investment and economic development*. UCL Press: London.

Bengston, D. N., Potts, R. S., Fan, D. P., and Goetz, E. G., (2005). An analysis of the public discourse about urban sprawl in the United States: Monitoring concern about a major threat to forests. *Forest Policy and Economics*, 7(5), pp. 745–756.

Berechman, J., and Paaswell, R. (2001). Accessibility improvements and local employment: An empirical analysis. *Journal of Transportation and Statistics,* 4(2–3), pp. 49–66.

Berion, P. (1998). Analyser les mobilités et le rayonnement des villes pour révéler les effets territoriaux des grandes infrastructures de transport. *Les Cahiers Scientifiques du Transport*, 33, pp. 109–127. [Analyze the mobility and influence of cities to reveal the territorial effects of major transport infrastructures. *The Scientific Papers of the Transport*]

Bhat, C., Handy, S., Kockelman, K., Mahmassani, H., Chen, Q. and Weston, L. (2000). *Development of an Urban Accessibility Index: Literature Review*, Research project conducted for the Texas Department of Transportation, Austin, USA; Center for Transportation Research, University of Texas, 92 p.

Battha, B. (2009). Modelling urban growth boundary using geoinformatics. *International Journal of Digital Earth*, 2(4), pp. 359-381.

Bhatta, B. (2010). *Analysis of urban growth and sprawl from remote sensing data*, New York, US, Springer, 172 p.

Breheny, M. (1997). Urban compaction: feasible and acceptable? *Cities*, 14(4), pp. 209-217.

Bruekner, J. K. and Fansler, D. A. (1983). The economics of urban sprawl: Theory and evidence on the spatial sizes of cities. *The Review of Economics and Statistics*, 65(3), pp. 479–482.

Bruekner, J. K. and. Largey, A. G. (2008). Social interaction and urban sprawl. *Journal of Urban Economics,* 64, pp. 18–34.

Buhalis, D., Eichhorn, V., Michopoulou, E. and Miller, G. (2005). *Accessibility market and stakeholder analysis*. U.K.: University of Surrey.

Burmeister, A., and Joignaux, G. (1997). *Infrastructures de transport et territoires: Approches de quelques grands projets*. Paris, l'Harmattan. [*Transport infrastructure and territories: Approaches to some major projects.*]

Camagni, R. (2005). *Economía urbana*, Barcelona, Antoni Bosch. [*Urban economy*]

Caves, D. W., Christensen, L. R. and Swanson, J. A. (1980). Productivity in US railroads 1951-75. *Bell Journal of Economics and Management Science*, 11, pp. 166-181.

Caves, D. W., Christensen, L. R. and Swanson, J. A. (1981). Economic performance in regulated and unregulated environments: A comparison of US and Canadian railroads. *Quarterly Journal of Economics*, 11, pp. 166-181.

Caves, D. W., Christensen, L. R., Swanson, J. A. and Tretheway, M. (1982). *Economic performance of US and Canadian railroads: The significance of ownership and regulatory environment.* In: Stanbury, W. T. and Thompson, F. (eds.), Praeger, New York, Managing Public Enterprise.

Cerda-Troncoso, J. and Marmolejo-Duarte, C. (2010). De la accesibilidad a la funcionalidad del territorio: una nueva dimensión para entender la estructura urbano-residencial de las áreas metropolitanas de Santiago (Chile) y Barcelona (España). *Revista de Geografía Norte Grande*, 46, pp. 5-27. [Of the accessibility to the functionality of the territory: a new dimension to understand the urban-residential structure of the metropolitan areas of Santiago (Chile) and Barcelona (Spain). *Geography Magazine Norte Grande*]

Cervero, R. (1999). Tracking accessibility: Employment and housing opportunities in the San Francisco Bay area. *Environmental Planning A*, 31(7), pp. 1259–1278.

Cervero, R. (2003). Road expansion, urban growth, and induced travel: A path analysis. *Journal of the American Planning Association*, 69(2), pp. 145–162.

Clark, D. (1982). *Urban Geography: An Introductory Guide.* London, Taylor & Francis, p. 231.

Coray D. and Manoj J. (2011). A dynamic modeling approach to investigate impacts to protected and low-income populations in highway planning. *Transportation Research Part A: Policy and Practice*, 45(7), pp. 598–610.

Correa-Díaz, G. (2010). Transporte y ciudad. *EURE*, 36(107), pp. 133-137. [Transport and city.]

Couch, C. and Karecha, J. (2006). Controlling urban sprawl: Some experiences from Liverpool. *Cities*, 23(5), pp. 353–363.

Currie, G., Richardson, T., Smyth, P., Vella-Broderick, D., Hine, J., Lucas, K., Stanley, J., Morris, J., Kinnear, R. and Stanley, J. (2009). Investigating links between transport disadvantage, social exclusion and well-being in Melbourne – Preliminary Results. *Transport Policy*, Special Issue International Perspectives on Transport and Social Exclusion, 16(3), pp. 90-96.

Daly, A. and Sillparchan, P. (2008). National Models. In: David Henser and Kenneth Button (eds.) *Handbook of Transport Modelling* (2nd edition), The Netherlands, Elsevier, pp. 489-502.

Duany, A., Plater-Zyberk, E., and Speck, J. (2001). *Suburban nation: The rise of sprawl and the decline of the American dream.* New York, North Point.

Dubois-Taine, G. (1991). *Les boulevards urbains, contribution à une politique de la ville.* Paris, Presses de l'Ecole Nationale des Ponts et Chaussées. [*Urban boulevards, contribution to a city policy.* Paris, Presses of the National School of Bridges and Roads]

Escolano, S. and Ortíz, J. (2005). La formación de un modelo policéntrico de la actividad comercial en el Gran Santiago (Chile). *Revista de Geografía Norte Grande*, 34, pp. 53-64. [The formation of a polycentric model of commercial activity in Greater Santiago (Chile). *Geography Magazine Norte Grande*]

Farhad, A. (1996). Reorienting metropolitan land use and transportation policies in the US. *Land Use Policy*, 13(1), pp. 37-49.

Fernández, S. (2000). ¿Qué se entiende por diseño universal?. *Entre dos mundos*, 13, pp. 21-26. [What is meant by universal design?. *Between two worlds*]

Forsyth, P. (2008). Models of Airport Performance. In: D. Henser and K. Button (eds.) *Handbook of Transport Modelling* (2nd edition). The Netherlands, Elsevier, pp. 715-727.

Friesz, T. and Kwon, C. (2008). Strategic Freight Network Planning Models and Dynamic Oligopolistic Urban Freight Networks. In: D. Henser and K. Button (eds.) *Handbook of Transport Modelling* (2nd edition). The Netherlands, Elsevier, pp. 611-631.

Friesz, T. L., Suo, Z. G. and Westin, L. (1998). Integration of freight network and computable general equilibrium models. In: L. Lundquist, L. G. Mattson, and T. J. Kim (eds.), *Network Infrastructure and the Urban Environment.* Berlin, Springer-Verlag.

Fuentes, C. (2009). La estructura espacial urbana y accesibilidad diferenciada a centros de empleo en Ciudad Juárez, Chihuahua. *Región y Sociedad*, 21(44), pp. 117-144. [The urban spatial structure and accessibility differentiated to employment centers in Ciudad Juárez, Chihuahua. *Region and Society*]

Fujita, M. and Thisse, J. (1996). Economics of Agglomerations, *Journal of the Japanese and international economies,* 10, pp. 339-378.

Fujita, M., Krugman, P. and Venables, A. (1999). *The Spatial Economy, Cities, Regions and International Trade.* Cambridge, MA, The MIT Press.

Gakenheimer, R. (1998). Los problemas de la movilidad en el mundo en desarrollo. *EURE*, 24(72), http://dx.doi.org/10.4067/S025071611998 007200002. [The problems of mobility in the developing world.]

Gallo-Rivera, M. T., Garrido-Yserte, R. and Vivar-Águila, M. (2010). Cambios territoriales en la Comunidad de Madrid: policentrismo y dispersión. *EURE*, 36(107), pp. 5-26. [Territorial changes in the Community of Madrid: polycentrism and dispersion]

García, J. (2010). Urban Sprawl and Travel to Work: The Case of the Metropolitan Area of Madrid, *Journal of Transport Geography*, 18(2), pp. 197-213.

Garrocho-Rangel, C. F. and Campos-Alanís, J. (2006). Un indicador de accesibilidad a unidades de servicio clave para ciudades mexicanas: fundamentos, diseño y aplicación, *Economía, sociedad y territorio*, 6(22), pp. 349-397. [An indicator of accessibility to key service units for Mexican cities: fundamentals, design and application, *Economy, society and territory*]

Geurs, K. T. and Ritsema van Eck, J. R. (2001). *Accessibility measures: review and applications*, RIVM report 408505 006. Bilthoven, National Institute of Public Health and the Environment www.rivm.nl/bibliotheek/rapporten/408505006.html.

Goodall, B. (1987). *Dictionary of human geography*, Harmondsworth, Penguin.

Grengs, J. (2010). Job accessibility and the modal mismatch in Detroit. *Journal of Transport Geography*, 18(1), pp. 42–54.

Guillermo, A. (2004). *Procesos metropolitanos y grandes ciudades: Dinámicas recientes en México y otros países.* México, Universidad

Nacional Autónoma de México. [*Metropolitan processes and large cities: Recent dynamics in Mexico and other countries.*]

Gutiérrez, A. (2000). La producción del transporte público en la metrópolis de Buenos Aires. Cambios recientes y tendencias futuras. *EURE*, 26(77), http://dx.doi.org/10.4067/S025071612000007700005. [The production of public transport in the metropolis of Buenos Aires. Recent changes and future trends.]

Gutiérrez, J. and García, J. (2006). Movilidad por motivo de trabajo en la comunidad de Madrid, *Revista del Instituto de Estudios Económicos*, 1-2, pp. 223-256. [Mobility due to work in the community of Madrid, *Journal of the Institute of Economic Studies*]

Guy, C. M. (1983). The assessment of access to local shopping opportunities: a comparison of accessibility measures, *Environment and Planning B*, 10, pp. 219–238.

Hamidi, S. and Ewing, R. (2014). A longitudinal study of changes in urban sprawl between 2000 and 2010 in the United States, *Landscape and Urban Planning*, 128, pp. 72–82.

Handy, S. L. and Niemeier, D. A., (1997). Measuring accessibility: an exploration of issues and alternatives, *Environment and Planning A*, 29, pp. 1175–1194.

Handy, S. L. and Clifton, K. (2001). Evaluating neighborhood accessibility: possibilities and practicalities, *Journal of Transportation and Statistics*, 4(2/3), pp. 67-78.

Hanson, S. (1986). *The geography of urban transportation*. New York, Guilford.

Haralambides, H. E. (2008). Structure and Operations in the Liner Shipping Industry. In: D. Henser and K. Button (eds.) *Handbook of Transport Modelling* (2nd edition). The Netherlands, Elsevier, pp. 761-775.

Harris, B. (2001). Accessibility: concepts and applications. *Journal of transportation and statistics*, 4(2/3), pp. 15-30.

Henry, E. (1998). Regards sur la mobilité urbaine a Amerique latine. *Espaces et Societés*, 2, pp. 52-58. [Perspectives on Urban Mobility in Latin America. *Spaces and Societies*]

Hiernaux-Nicolas, D. (2005). ¿Identidades móviles o movilidad sin identidad? El individuo moderno en transformación. *Revista de Geografía Norte Grande*, 34, pp. 5-17. [Mobile identities or mobility without identity? The modern individual in transformation. *Geography Magazine Norte Grande*]

Holl, A. (2004a). Manufacturing location and impacts of road transport infrastructure: empirical evidence from Spain, *Regional Science and Urban Economics,* 34(3), pp. 341-363.

Holl, A. (2004b). Transport Infrastructure, Agglomeration Economies, and Firm Birth: Empirical Evidence from Portugal, *Journal of Regional Science*, 44(4), pp. 693-712.

Houston, D. (2012). *Getting to work: Spatial mismatches between the unemployed and jobs.*

Jara-Díaz, S. R. and Martínez, F. J. (1999). On the specification of indirect utility and willingness-to-pay for discrete residential location models. *Journal of Regional Science*, 39, pp. 675-688.

Johnson, C. M. (2009). Cross-border regions and territorial restructuring in central Europe room for more transboundary space, *European Urban and Regional Studies*, 16(2), pp. 177-191.

Jouffe, Y. and Lazo-Corvalán, A. (2010). Las prácticas cotidianas frente a los dispositivos de la movilidad. Aproximación política a la movilidad cotidiana de las poblaciones pobres periurbanas de Santiago de Chile. *EURE*, 36(108), pp. 29-47. [The daily practices in front of the mobility devices. Political approach to the daily mobility of the poor peri-urban populations of Santiago de Chile.]

Klug, S., and Yoshitsugu, H. (2012). Urban sprawl and local infrastructure in Japan and Germany. *Journal of Infrastructure Systems*, 10.1061/(ASCE)IS.1943-555X.0000101, 232–241.

Lenntorp, B. (1976). *Paths in Time-Space Environments: A Time Geographic Study of Movement Possibilities of individuals*, Lund Studies in Geography, B series, 44, Stockholm, Gleerup, pp. 150.

Lizarraga, C. (2012). Expansión metropolitana y movilidad: el caso de Caracas. *EURE*, 38(113), pp. 99-125. [Metropolitan expansion and mobility: the case of Caracas]

Malayath, M. and Verma, A. (2013). Activity based travel demand models as a tool for evaluating sustainable transportation policies. *Research in Transportation Economics*, 38(1), pp. 45–66.

Martínez, F. (1992). The bid-choice land use model: An integrated economic framework. *Environment and Planning A*, 24, pp. 871-885.

Martínez, F. (2008). Towards a Land-Use and Transport Interaction Framework. In: D. Henser and K. Button (eds.) *Handbook of Transport Modelling* (2nd edition). The Netherlands, Elsevier, pp. 181-201.

McFadden, D. (1974). Conditional logit analysis of qualitative choice behavior. In: P. Zarembka (ed.), *Frontiers in Econometrics*, New York, Academic Press, pp. 105-142.

McFadden, D. (1978). Modelling the choice of residential location. In: Karlqvist, A., Lundqvist, L., Snickars, F. and Weibull, J. W. (eds.), *Spatial Interaction Theory and Planning Models*, Amsterdam, North-Holland, pp. 75-96.

McGranahan, G., Satterthwaite, D., and Tacoli, C. (2004). *Rural-urban change boundary problems and environmental burdens*. London, International Institute for Environment and Development, 25 p.

McNally, M. (2008). The Four Step Model. In: D. Henser and K. Button (eds.) *Handbook of Transport Modelling* (2nd edition), The Netherlands, Elsevier, pp. 35-53.

Mendoza-Cota, J. and Pérez-Cruz, J. (2007). Aglomeración, encadenamientos industriales y cambios de localización manufacturera en México, *Economía, Sociedad y Territorio*, VI(23), pp. 655-691. [Agglomeration, industrial linkages and changes in manufacturing location in Mexico, *Economy, Society and Territory*]

Miller, H. J. and Wu, Y.-H. (2000). GIS Software for Measuring Space-Time Accessibility in Transportation Planning and Analysis, *GeoInformatica*, 4, pp. 141-159.

Millward, H. and Spinney, J. (2011). Time use, travel behavior, and the rural–urban continuum: Results from the Halifax STAR Project. *Journal of Transport Geography*, 19(1), pp. 51–58.

Miralles-Guasch, C. (2012) Las encuestas de movilidad y los referentes ambientales de los transportes. *EURE*, 38(115), pp. 33-45. [The mobility surveys and the environmental references of transport.]

Modarres, A. (2013). Commuting and energy consumption: toward an equitable transportation policy. *Journal of Transport Geography*, 33, pp. 240-249.

Morris, J. M., Dumble, P. L. and Ramsay M. W. (1979). Accessibility Indicators for Transport Planning. *Transportation Research A*, 13, pp. 91-109.

Murakami, E. and Wagner, D. (1999). Can using global positioning system GPS improve trip reporting? *Transportation Research Part C: Emerging Technologies*, 7(2–3), pp. 149–165.

Nash, C. and Smith, A. (2008). Modelling Performance: Rail. In: D. Henser and K. Button (eds.), *Handbook of Transport Modelling* (2nd edition). The Netherlands, Elsevier, pp. 665-691.

Newman, P., Kenworthy, J. and Vintila, P. (1995). Can we overcome automobile dependence? Physical planning in an age of urban cynicism. *Cities*, 12(1), pp. 53-65.

Nutley, S. (2003). Indicators of transport and accessibility problems in rural Australia. *Journal of Transport Geography*, 11(1), pp. 55–71.

Obregón-Biosca, S. (2010). Estudio comparativo del impacto en el desarrollo socioeconómico en dos carreteras: Eix Transversal de Catalunya, España y MEX120, México. *Economía Sociedad y Territorio*, 10(32), pp. 1-47. [Comparative study of the impact on socioeconomic development on two roads: Eix Transversal de Catalunya, Spain and MEX120, Mexico. *Economy Society and Territory*]

Obregón-Biosca, S. and Junyent, R. (2011). The Socioeconomic Impact of the Roads: A Case Study of the 'Eix Transversal' in Catalonia, Spain. *Journal of Urban Planning and Development*, 137(2), pp. 159–170.

Obregón-Biosca, S., Sánchez-Escobedo, J. A. and Somohano-Martínez, Ma. L. (2016a). Planificación de rutas turísticas para autobús a través de indicadores de accesibilidad integral y de dotación de bienes materiales e inmateriales. *Revista Transporte y Territorio*, 14, pp. 144-

166. [Planning of tourist routes for bus through indicators of integral accessibility and endowment of material and immaterial goods. *Transport and Territory Magazine*]

Obregón-Biosca, S., Romero, J. A., Mendoza, J. F. and Betanzo, E. (2016b). Impact of Mobility Induced by Urban Sprawl: Case Study of the Querétaro Metropolitan Area. *Journal of Urban Planning and Development*, 142(2), 05015005.

Oppenheim, N. (1995). *Urban Travel Demand Modelling*. New York, John Wiley and Sons, Inc.

Ortíz, J. and Escolano, S. (2013). Movilidad residencial del sector de renta alta del Gran Santiago (Chile): hacia el aumento de la complejidad de los patrones socioespaciales de segregación. *EURE*, 39(118), pp. 77-96. [Residential mobility of the high income sector of Greater Santiago (Chile): towards increasing the complexity of socio-spatial patterns of segregation]

Ortúzar, J. D. and Willumsen, L. (2008). *Transport models*, Spain, Universidad de Cantabria, Cantalabria.

Oum, T. H., Waters II, W. G. and Yu, C. (1999). A survey of productivity and efficiency measures in rail transport. *Journal of Transport Economics and Policy*, 33, pp. 9-42.

Ozbay, K., Ozmen, D. and Berechman, J. (2003). Empirical analysis of relationship between accessibility and economic development. *Journal of Urban Planning and Development*, 129(2), pp. 97–119.

Ozbay, K., Ozmen, D. and Berechman, J. (2006). Modeling and analysis of the link between accessibility and employment growth. *Journal of Transportation Engineering*, 132(5), pp. 385-393.

Ozmen, D., Ozbay, K., and Holguin-Veras, J. (2007). Role of transportation accessibility in attracting new businesses to New Jersey. *Journal of Urban Planning and Development*, 133(2), pp. 138–149.

Paaswell, R. and Zupan, J. (2007). *Transportation Infrastructure Investments: New York and its Global Peers*, Working paper, New York, City College of New York.

Pardo, C. (2005). Salida de emergencia: reflexiones sociales sobre las políticas del transporte. *Universitas Psychologica*, 4(3), pp. 271-284.

[Emergency exit: social reflections on transport policies. *Psychological University*]

Pengjun, Z. (2013). Too complex to be managed? New trends in peri-urbanisation and its planning in Beijing. *Cities*, 30, pp. 68–76.

Plassard, F. (1978). *Les autoroutes et le développement regional*. Paris, Pul-Economica. [*Highways and regional development*]

Putnam, R. D. (2000). *Bowling Alone*, New York, Simon and Schuster.

Quayle, M. and Driessen Van der Lieck, T. (1997). Growing community: A case for hybrid landscapes. *Landscape and Urban Planning*, 39(2), pp. 99–107.

Rahman, G., Alam, D. and Islam, S. (2008). City growth with urban sprawl and problems of management, in *Proceedings of 44th ISOCARP Congress Dalian*, China, September 19 -23, International Society of City and Regional Planners and Urban Planning Society of China.

Ríos-Quezada, G. and Obregón-Biosca, S. (2017). La accesibilidad de las autovías y la teoría de localización industrial. *Economía, Sociedad y Territorio*, 17(55), pp. 581-617. [The accessibility of the highways and the theory of industrial location. *Economy, Society and Territory*]

Rodríguez, J. (2012). ¿Policentrismo o ampliación de la centralidad histórica en el Área Metropolitana del Gran Santiago? Evidencia novedosa proveniente de la encuesta Casen 2009. *EURE*, 38(114), pp. 71-97. [Polycentrism or expansion of historical centrality in the Metropolitan Area of Greater Santiago? New evidence from the Casen 2009 survey]

Rouse, P. and Putterill, M. (2008). Highway Performance. In: D. Henser and K. Button (eds.), *Handbook of Transport Modelling* (2nd edition). The Netherlands, Elsevier, pp. 743-760.

Ruiz-Juri, N. and Kockelman, K. (2006). Evaluation of the Trans-Texas corridor proposal: Application and enhancements of the random-utility based multiregional input–output model. *Journal of Transportation Engineering*, 132(7), pp. 531–539.

Ryan, S. (1999). Property values and transportation facilities: finding the transportation-land use connection. *Journal of Planning Literature*, 13(4), pp. 412-427.

Sanchez-Reaza, J. (2010). Trade, Proximity and Gwowth: The Impact of Economic Integration on Mexico's Regional Disparities. *Integration & Trade Journal*, 14(31), pp. 19-32.

Sanchez, T. W. (1999). The connection between public transit and employment: The cases of Portland and Atlanta. *Journal of American Planning Association*, 65(3), pp. 284–296.

Schmitta, B. and Henry, M. S. (2000). Size and growth of urban centers in French labor market areas: consequences for rural population and employment. *Regional Science and Urban Economics*, 30(1), pp. 1–21.

Siccardi, A. J. (1986). Economic effects of transit and highway construction and rehabilitation. *Journal of Transportation Engineering*, 112(1), pp. 63–76.

Smith, A. S. J. (2006). Are Britanian's railways costing too much? Perspectives base don TFP comparisons with British rail; 1963-2002. *Journal of Transport Economics and Policy*, 40, pp. 1-45.

Sobrino, J. (1993). *Gobierno y administración metropolitana y regional*. México, Instituto Nacional de Administración Pública. [*Government and metropolitan and regional administration*. Mexico, National Institute of Public Administration]

Sobrino, J. (2007). Standards for intra-metropolitan dispersion in México. *Estudios Demograficos Urbanos*, 22(3), pp. 583–617. [*Urban Demographic Studies*]

Song, S. (1992). *Monocentric and Polycentric Density Functions and their Required Commutes*. Working paper UCTC 198, University of California Transportation Center.

Soto-Caro, M. and Álvarez-Aránguiz, L. (2012). Análisis de tendencias en movilidad en el Gran Valparaíso. El caso de la movilidad laboral. *Revista de Geografía Norte Grande*, 52, pp. 19-36. [Analysis of trends in mobility in the Greater Valparaíso. The case of labor mobility. *Geography Magazine Norte Grande*]

Stopher, P., Fitzgerald, C. and Xu, M. (2007). Assessing the accuracy of the Sydney Household Travel Survey with GPS. *Transportation*, 34(6), pp. 723–741.

Suárez-Lastra, M. and Delgado-Campos, J. (2010). Patrones de movilidad residencial en la Ciudad de México como evidencia de co-localización de población y empleos. *EURE*, 36(107) pp. 67-91. [Patterns of residential mobility in Mexico City as evidence of co-location of population and jobs.]

Sudhira, H. S., and Ramachandra, T. V. (2007). Characterising urban sprawl from remote sensing data using landscape metrics. *Proceedings of the 10th International Conference on Computers in Urban Planning and Urban Management*, Conselho Nacional de Desenvolvimento Científico e Tecnológico, Brasillia, Brazil.

Sutton, C. (1999). Land use change along Denver's I-225 beltway. *Journal of Transportation Geography*, 7(1), pp. 31–41.

Thakuriah, P. (2001). Introduction to the Special Issue on Methodological Issues in Accessibility Measures with Possible Policy Implications. *Journal of Transportation and Statistics*, 4(2/3) pp. V.

Ureta-Icaza, S. (2009). Manejando por Santiago. Explorando el uso de automóviles por parte de habitantes de bajos ingresos desde una óptica de movilidad sustentable. *EURE*, 35(105), pp. 71-93. [Driving through Santiago. Exploring the use of automobiles by low-income inhabitants from a perspective of sustainable mobility.]

Valero, Á. (1984). Movilidad espacial en Madrid. *Anales de Geografía de la Universidad Complutense*, 4, pp. 207-225. [Space mobility in Madrid. *Annals of Geography of the Complutense University*]

Weibull, J. W. (1976). An axiomatic approach to the measurement of accessibility. *Regional Science and Urban Economics*, 6, pp. 357–379.

Willumsen, L. G. (2008). Travel Networks. In: D. Henser and K. Button (eds.) *Handbook of Transport Modelling* (2nd edition). The Netherlands, Elsevier, pp. 203-220.

Zhang, B. (2004). *Study on urban growth management in China*, Beijing, Xinhua Press. pp. 235.

In: Transportation Infrastructure ISBN: 978-1-53614-059-0
Editor: S. Antonio Obregón Biosca © 2018 Nova Science Publishers, Inc.

Chapter 2

THE EXPERIENCE OF ACTIVE MOBILITY AND ITS CONTRIBUTIONS TO URBAN HABITABILITY

Avatar Flores-Gutiérrez[*], *PhD*,
Guillermo I. López-Domínguez, MD
and Verónica Leyva-Picazo, MD

Graduate and Undergraduate Program of Architecture,
Faculty of Engineering, Universidad Autónoma de Querétaro (UAQ),
Queretaro, Queretaro, Mexico

ABSTRACT

For humans, mobility in urban realms represents a means of interaction with other members of the community. The types of transportation chosen are, to a large extent, a reflection of a personal and community idiosyncrasy. Active mobility is undoubtedly the form of transportation that most contributes to contact among the inhabitants of a city by maintaining face-to-face interactions. This chapter discusses the

---

[*] Corresponding Author Email: avatar.flores@uaq.mx.

concept of urban habitability and the importance of specifically designing infrastructure for active mobility, considering its contributions and potential for more livable cities from the point of view of its inhabitants.

Keywords: active mobility, habitability, urban habitability, sense of community

INTRODUCTION

Humans, as social beings, form habitats in a conglomerate of houses, factories and other spaces, all of them connected by roads. The objective is simple: the city provides them with a space for the exchange of goods and services, but, above all, it gives them security and increases their chances of survival.

Our ecology is social, according to Lopez-Riquelme (2018). Ensuring our survival depends to a large extent on human relationships between the individuals that make up a society. Even from an evolutionary perspective, human behavior has a specific purpose, reflected in the social life of an urban community: survival. The quality of a community depends on the social framework that is updated on a daily basis.

In this chapter we review the relationship between mobility and urban habitability. Specifically, we discuss the impact that mobility may have in strengthening social ties and providing the means to attain common goals. In the first part of the chapter, we start by understanding mobility as a social manifestation that reflects the idiosyncrasy of a particular society. This can be recognized in the way different perspectives have shaped cities, reflecting specific paradigms. In the second part, we elaborate on the concept of habitability in reference to an urban context. We claim that active mobility is an essential element in the relationship of habitability and the means of transport. This includes the experience of living and commuting through active mobility on the one hand, and the impact of having the automobile as the privileged form of transport in major cities, on the other. The chapter closes discussing the importance of designing specific infrastructure for active mobility, emphasizing urban habitability and its potential as a social bond.

PART I: URBAN MOBILITY AND IDIOSYNCRASY

Expectations of Mobility Are Only a Reflection of Society

At first glance, the central aim of mobility seems evident – getting people from one place to another, connecting the different meaningful sites of a regular citizen. But, along with that objective, within the city there is a set of byproducts, both positive and negative, such as traffic, pollution, noise, and also a broader range for interaction. People are no longer limited to a few hundred meters surrounding their neighborhood; as they move faster, the city is perceived as being smaller.

Mobility appears as the central node to this cluster of related activities. It is often assumed to belong to a set of issues, which need to be addressed in big cities, where the density makes it evident. But what would happen if for the sake of analysis, we assume that mobility is not at the core of the cluster, but rather consider it as a material and tangible manifestation of certain social realities? In this section, we explore the possible connections between identity and idiosyncrasy that may be behind the mobility choices of citizens.

The selection of a particular means of transportation has a lot to do with availability and economy. Nonetheless, it has a cultural component. A big part of this selection seems to be predetermined from the beginning as individuals in a society, consciously or unconsciously, try to figure out which type of mobility is appropriate for them. Socially, some societies find that bicycles are better suited for young people, just as families assume to need SUVs for their mobility. Often these decisions are not made entirely objectively. Instead, they are based on what other members of a community do. Carmakers (and for that matter, almost any industry trying to sell something) are continuously profiling their potential consumers to find out what their tastes are and finding ways to appeal to these expectations. It is also true, that in some other cases, it is mainly the geographical conditions or limited availability that are the main reasons for making a mobility choice.

Individualist Societies in Which Mobility Choices Are a Mere Reflection of This Condition

Paradoxically, contemporary societies tend to emphasize the idea of competition over cooperation. This is particularly true in societies with a big economic gap between different socio-economic layers. This variation often encourages the need for differentiation among social classes as well as the perceived necessity to climb the socio-economic ladder. Private transport has long been perceived as representing the status of social mobility. Vehicles are a commodity that can be taken along and one that is visible to others. A good part of the population is eager to get loans and buy cars, instead of making more durable or reliable investments. Undeniably, this is partly due to the need for transportation. Yet, this necessity is accompanied by a desire for differentiation and status, even if these vehicles have a relatively short life-span and need to be renewed frequently to keep up. Private cars represent a visible condition, a reasonable expectation to invest in, particularly because beyond the explicit purpose of transportation, these also provide a new status. This is one of the reasons why private transport is often preferred as an investment over housing.

On the other hand, regardless where in the socio-economic scale they are located (poor, rich or middle class) societies with smaller socio-economic gaps tend to be more cooperative or at least more prone to consider collective mobility. One of the reasons behind this is that since most of the individuals have similar incomes and properties, there is no need to demonstrate something or to invest in private vehicles as a means of status, as would happen in unequal societies. Suddenly, transportation is deprived of most of the constructs around it, and as it gets depleted of such types of meanings, it becomes more widespread and no longer symbolic.

And then, the issue of mobility can in turn be dealt with in more objective terms.

Still, it must be added, that society and individuals can be approached, not only from the economic and symbolic points of view, but also by creating a new identity or, at least, a new identification with certain alternatives. This cultural shift not only needs to be worked out and planned in order to succeed, but it is even better when this new identity comes from within the same society – a process in which its members may be involved and in which the public means of transportation can be perceived as an accomplishment and not only as a random consequence of public policies or private interests. In that sense, along with public space, mobility can become not just a source of identity but even something more complex, functioning as a balance that provides a certain equilibrium in societies.

Public Space and Mobility as a Compensating Mechanism

Mobility represents a large part of the budget for families, according to Vasconcelos (2011) in *Desarrollo urbano y movilidad en América Latina*. The relationship between public and private transport can range from twice the cost up to eight times that of private transport over public. This means that while a public ride costs about 0.70 USD, the same distance costs about 4 USD by private transport. And it must be noted that a large part of a commuter´s budget is basically used to pay for commuting from home to work and back. Thus, a lot of the earned money is spent just in getting to the same job that provides the income for these necessary, basic transportation expenses.

If we add to this, the constant expansion of cities via peripheral neighborhoods and the uneven distribution of income, it would almost be possible to track how people are traveling between different parts of the cities, all depending on the given geographical and economic information.

Besides, marginalization is usually emphasized by a lack of infrastructure in mobility. Either few roads connect densely populated areas or paradoxically, wide roads are specifically designed for the private mobility of the richer neighborhoods. The morphology of the city is predisposed by these factors; and in consequence, the mobility options are narrowed from the beginning, which in turn means that different sectors in society initially assume that they have less options. Especially for societies in developing countries, this implies private transport for the richer social classes and a faulty public transport for lower income families.

This situation implies that if public and private institutions take the lead and shift the perception from individual to social advances, mobility has the potential to become a symbol of improvement. While the economic gap in societies should be targeted in different ways to be effectively reduced, there are other mechanisms that have a huge impact on the perception of the quality of life in populations, such as mobility and public space.

In cities, like Medellín, Colombia, with a unique topography, surrounded by steep hills makes for many of the marginalized neighborhoods to sit on the mountain slopes. People from these places would spend up to four hours commuting every day from their home to their job. After the funicular was installed and their commuting times cut in half, their perception of themselves changed positively (Personal Interview with Jeihhco, Jeisson Alexander Castaño, Hopper and the cultural manager at *Comuna 13*, Medellín, June 2012). Even if their economic situation had not improved, at least they became aware that they had more spare time to spend with their families. They also had a slight advantage over the upper classes, since these were still dependent on using private cars. It is in this renewed perception, along with the creation of a new identity or at least, a sort of appropriation of the new transport system, that the private transport tendency is reverted or compensated.

Still the value of time – specifically personal and family time – is probably underestimated, particularly in societies in which *free time* is interpreted as laziness. Yet several developed countries have policies that encourage free time for employees and workers; this in turn provides other needs, but specifically, other possibilities for mobility, especially when people are not in a rush and can be creative to seek alternatives.

The other aspect that helps improve this social image and change the individual idiosyncrasy to a more cooperative, safe environment, which in turn improves habitability, is public space. Even if houses are small or not well equipped (this applies for to poor neighborhoods as well as for any densely populated area, where there are heavy space restrictions), the perception of a good quality of life and in turn, an improved habitability, can be attained with a rich and lively public space.

Public space, by improving the perception of safety and opening the opportunity of several outdoor activities, fosters the experience of walking around neighborhoods to run small errands. In many mild-weathered cities, even if the natural conditions allow a regular outdoor experience, this is usually avoided, because of the physical or unsafe conditions.

People tend to feel compensated living in a neighborhood that provides a safe and comfortable public space, especially when it has a varied program along throughout the day. This public space turns into an extension of their homes. This means that not all of their lives have to occur inside a small house, hidden (and protected) from the exterior, but instead a lot of their life can be developed in a context with increased social interaction, and maybe, above all, in an increased awareness of their surroundings. This influences the perception of distance, such that walking or cycling become more common, viable alternatives over vehicles.

Mobility Trends

Finally, there is another factor that needs to be taken into account, which involves cultural shifts and trends. As frivolous as it may seem, it does indeed reflect another reality, one that is frequently obviated while discussing mobility. Just as during the mid-part of the 20th Century, the car represented freedom, status and progress, while other types of mobility were regarded as inferior or rudimentary, a growing, awareness of environmental issues, as well as a potentially better informed generation, have begun thinking about other possibilities for mobility. Starting from cycling mobility, these obviously expand to other transportation means, including public transport and electric or hybrid vehicles. Additionally, some other trends, such as Neo-ruralism, home offices, freelancers, people into *organic* and *natural* lifestyles can all have an impact, hopefully positive, on mobility. It is true that even if this may sound romantic, not all of these are absolutely free or spontaneous – a lot of them will also rely on business models that would be accommodating these trends as they grow and consolidate. On the other hand, existing industries will also try to keep up business (including traditional private transport) with even more aggressive techniques.

This (or these) new generation(s) translates their sense of belonging into other cultural trends. Some of them can be temporary or non-transcendental. Some may be long lasting. It can be argued that these things, just keep the status symbol alive, just changing the emphasis from the luxury of a car, for example, to the *virtue* of ecology or to alternative energy. But in any case, all of them also interpret the needs and expectations of current times.

It is true that at this point, these trends may not be a large enough sample to measure the impact on overall mobility. But if we consider that trends like this one tend to behave in a Gauss curve (Erner 2013), it is very likely that sooner or later, early adopters will be followed by a larger population, if not immediately, probably in upcoming generations; and certainly both at public and private levels, this will create a renewed market for different types of investments.

Another topic to consider is that mobility is not only a result of these cultural factors that have been discussed so far. Mobility is also a directed and encouraged (or discouraged) activity by public institutions and policies, as well as by private enterprises. For instance, governments have actively sought to get private investors to create jobs. Motor vehicles are a big part of this type of industries with the capability of creating a lot of direct and indirect jobs. Such factories are often subsidized by governments, eminently interested in providing an appropriate market (even if a major part of the production is exported). This, along with policies focused on building roads (which are very visible public investments), pushes further the idea that private transport should prevail.

On the other hand, investment in public transport has declined over the years in several places. In several countries, subway systems and railroads (examples of collective transport) have decreased or at least are not as dynamic as they should be, because of the associated immediate costs and the major cultural shift that needs to be attained.

Another political reason is that government administrations are actively seeking foreign investment. In places such a central México, a lot of this investment is related to the assembly of cars, different types of vehicles, and the production of their components. Even if much of this production is done for export, a lot of it still impacts in the local market in two ways: directly by injecting goods (in this case vehicles) and secondly, with more job offers and manufacturing plants in the outskirts of cities (meaning an increased need in transportation for hundreds if not thousands of workers).

Far beyond the general need for transportation than can be traced in the necessity for regular commutes between job, home and leisure and which can somehow be quantified even if mobility is deeply influenced by expectations, culture and idiosyncrasy, the choice or preference in mobility can also be adjusted, turned off, or modified. Public and private interests are at stake when promoting alternatives, but often, the common good is postponed in favor of other economical or political goals. Still, when thinking in the long term, all these stakeholders could benefit in promoting active mobility and reducing the economic gap in cities.

PART II: HABITABILITY AND THE EXPERIENCE OF MOBILITY IN THE URBAN ENVIRONMENT

The Concept of Habitability in Cities

Habitability is a concept whereby its definition is usually circumscribed to the architectural field, or more specifically, to that of livable spaces. In commenting on cities, it is more common to use the term life quality. Still the concept of habitability encompasses a human dimension that when translated to the urban scale, becomes a social dimension. Let us first clear up the meaning of the term, before discussing urban habitability.

According to Flores-Gutiérrez (2016), habitability is the possibility that spaces provide human beings for realizing their complex activities with the aim of solving their necessities. When referring to *complex activities* Flores-Gutiérrez takes into account physical as well as psychological activities, something important to consider, as it will be further developed in this same section.

This definition is wide enough to translate it into the urban dimension, by substituting the word *spaces* for the word *environment*, which covers the totality of the experience. Perhaps it is closer to the concept *atmosphere*, as narrated by Zumthor (2010) in his description of a public plaza as a place that can only be understood for its integration of experiences, such as sounds, temperature, material, textures, and of course even, the interaction of people.

Additionally, a definition of habitability should consider that an individual dimension of the human being is essential in conforming livable architectural space. On an urban level, it is impossible not to consider the community as that individual which the environment must address. Cities should be communities. In them a variety of desires and activities, as well as a plurality of thought, converge, but they must be thought under the same premise: community should be privileged over individuals.

Thus, in the urban realm, we stress that habitability is the possibility that environments provide to the community for realizing their complex activities with the objective of solving their common necessities.

Which are these necessities of the community? And which are the complex activities that must be done to address them?

First, let us remember that as Zumthor (2010), from architecture, or Gifford (2007), from environmental psychology, explain: A human being as an individual is in itself one of the essential components of the environment. The perception of the habitat is tremendously influenced by density and the behavior of other individuals, such that the way we perceive the cities depends not only on its streets, buildings or infrastructure, but also on the behavior of other individuals in shared spaces, which in turn are influenced by the environment itself.

As previously stated, privileging the collective is fundamental for the development of individuals in a community; nurturing a sense of belonging promotes empathy and collaboration among inhabitants. They, in turn, can collaborate toward the creation of more livable cities through their actions. The infrastructure cannot provide this sense of belonging on its own, but it can foster human interaction, which is, indeed, capable of creating this sense of belonging among people.

Attaining it would only be possible to the extent to which individuals feel related with each other, and to the extent in which they recognize themselves as inhabitants of the same environment. Physical and psychological human interaction is the activity required to solve the conditions for living in a community and creating a vital support for individuals.

Active Mobility as a Component of Urban Habitability

Individuals are related to each other in an urban environment; they are related publicly (and privately) through plazas, common spaces of interaction such as civic and cultural halls, roads, collective transport, and transitional spaces between destinations, which coincidentally are usually private.

The first group mentioned are spaces where there is a choice to visit them and where an interaction is expected. The second group pertains to spaces of transition and roads; these are places where there is an involuntary interaction, occurring almost every day in people's lives.

Gehl (2006) makes an additional distinction based on the type of activity involved, regarding needed and optional activities as well as (resultant) social activities. The latter, the social, are those to which we consider every time we think about public space. They are those activities that "depend on the presence of other people in public spaces (…), including playgrounds for kids, greetings, conversations, different sorts of common activities, and lastly -as the most extended social activity-, the contacts, passive in nature, such as seeing and listening other people" (Gehl, 2006).

Nevertheless, our topic for this book, *mobility*, occurs just in those transition spaces and in activities that, just as Gehl (2006) mentions, could be more of the needed type, that is, activities that

"are more or less mandatory, like going to school or to work, go shopping, wait for the bus or for a person, run errands or deliver the mail, in other words, all those activities in which the people involved are more or less forced to participate" (Gehl 2006).

These activities occur mainly in roads that connect individual destinations with private intentions. The passing of people through these "public spaces" occurs under different circumstances, mainly because of the decision and specific possibilities of the individuals, such that they can happen in a public, collective, or private manner.

The case of the automobile is the extreme of private mobility – an extension of the private space of the house, connected through a private "tunnel" (automobile) with another private space. As it has been stated in this text, nothing portrays more accurately the individuality of a culture than its choice for private transportation, as the automobile. It happens in the more individualistic societies with a diminished sense of community. The contact with the rest of the members of the community is almost non-existent, while isolated in a climatized and sonorized module that travels interacting only with other lifeless "motor beings," which are obstacles on a road that they would rather not take.

On the opposite end, there is individual and collective mobility in which there is a direct interaction with community – public transport, bicycle or walking, for instance. An important component of this type of mobility is that it involves physical force from people, and that is known as *active mobility*.

A lot has been said about the benefits of active mobility for societies that privilege it and it is currently admitted that active mobility, mainly walking and cycling, substantially contributes "to the population's physical and mental health" (Hess et al. 2017). Nevertheless, even if less studied, it is of utmost importance to pay attention to the enormous potential of active mobility as a transforming agent in social interaction at an urban level. The benefits of inhabiting cities where a strong sense of belonging is present and as a consequence, respect and equality among its dwellers, is commented at the beginning of this chapter.

Encompassed in this vision of mobility are the resultant activities to which Gehl (2006) refers; those that "derive of activities linked (…) to the categories of mandatory and social activities (…), because people are in the same space, they meet, cross or are just at sight."

Active mobility, then, puts human beings in public open spaces, where they will certainly interact with other human beings, promoting human contact. This space, when conceived with quality, "provides an informal, relaxed means of maintaining social ties and a sense of community in an urban environment" (Gehl 2010, quoted in Church 2012). This will be further discussed further on.

For example, let us consider the experience of riding a bicycle or of a walking journey. The possibility of inhabiting at all times the different spaces that are traversed is one of the most remarkable characteristics of active mobility. When walking, for instance, one has a broader perception of reality. Is it that what we see through the windshield in a car is not reality? It certainly is, but reality, the magic of reality, as Zumthor calls it, includes environmental dimensions that include not only the visual, but many other factors like temperature, wind, textures in pavement, and the materials that make up the city, and of course, the perception we have of other people thought the murmur of their voices and their actions – a more complex and multi-sensory reality than perceived through a car window.

According to Gifford (2007), "In the environmental perception, the perception of large scale scenarios as an entity is emphasized. Who perceives is also part of the scene and observes from multiple perspectives" (Gifford 2007). It follows then, that the perception of a space is real as long as it involves the totality of the environment with all its components.

Besides the great difference implied in the total perception of the environment through journeys in a car, by foot or bicycle, there is yet another distinction – the possibility to stop, look around, or change route. It is difficult for a car to make a stop in the middle of the street and take a glance at something that calls for attention. It is even more difficult to even find something worthy of attention outside of the narrow field of vision through the windshield or the rearview mirrors thought which drivers see onto the asphalt.

> "It is symptomatic, that in general, both modern cities and its dwellers underestimate the importance of walking. For them, it is only an inefficient means of transport, nonetheless, walking is emblematic as a primary way of perception." (Pérez-Gómez 2017)

By foot, the world scale changes. The odor of coffee or of a delicious breakfast can prompt a pause on the way. The sound of children out of school will promote a glance, memories and reflections. A reflection on a window will make us look aside; the feeling of the sun or a cool shadow will make us change sidewalk. Along with all the encounters and mismatches, these behaviors can provoke our interactions with the city and with people.

The habitability of the city, of which we have previously spoken, happens fostered by active mobility. Encounter, empathy, sense of community and belonging are essential aspects of human beings and are required components for habitability in the city.

If the evidence shows the potential for cities about a different way of inhabiting and traveling through them on a human scale, it is worth wondering why these forms of mobility are the least chosen in many big cities.

According to Pérez-Gómez (2017), the problem lies in the changes in the paradigm that has accompanied the city:

> "Urban planners prevailed over architects and urban designers, adopting the values of engineers serving political and economic interests; the reason, utility and efficiency became the determinant factors of a physical environment that was gradually getting sterile (…) Cities were designed by the "calculation" of rational planners that were naturally giving too little importance to the atmosphere of public space, focusing instead on urban networks that could efficiently accommodate traffic; spaces that could easily be observed and controlled thanks to their neutral character."

When the private, motorized vehicle is the paradigm, the whole city is transformed and adapted to this vision: government decisions and the behavior of drivers, even cyclists and pedestrians, understand their position in the hierarchy of mobility, in a vision where what matters is that vehicles travel without pausing and with the least possible obstacles (people!).

The current paradigm, not only is not stimulating interactions, but it prevents any attempt in society to explore different ways of mobility. First of all, due to a lack of infrastructure, insecurity, or design mistakes, pedestrians and cyclists are at risk in a world made for cars. Particularly outstanding is the mistake of not understanding that the goal is not to create more sidewalks or cycle paths, but the habitability of the city, which implies another way of designing, providing opportunities beyond riding and walking surfaces.

A change of paradigm is urgent; it is already happening in a lot of cities, especially in Europe, where the urban hierarchies of mobility start to make sense: people first, wheels second. Designing cities in which their paradigm allows habitability, requires, as mentioned, an understanding that it is not enough to provide minimum infrastructure, but that the infrastructure should be perceived as desired and it be capable of fostering human activity according to the urban offer.

The Perception of Infrastructure for Active Mobility

Designing infrastructure for active mobility is fundamental for making a change in the paradigm that could transform cities into more livable environments.

Any planner, on first impulse, will suspect that the tools of urban infrastructure, such as streets, sidewalks, roads, lighting or signaling, have little to do with people's behavior, and that they will not be enough to change the paradigm in cities. This is partly true, as long as it does not comply with the potential of *atmospheres* to influence human behavior.

> "Even though the physical frame does not have a direct influence on quality, content and intensity of social contacts, architects and urbanists can influence the possibilities of finding, seeing and listening to other people, possibilities that imply a quality in itself and that become important as a background and starting point for other forms of contact." (Gehl 2006)

Gehl's ideas are not isolated observations and have an explanation in the Affordances Theory by environment psychologist J. J. Gibson:

> "Affordances are preconditions for activity (…). The presence in a situation of a system that provides an affordance for some activity does not imply that the activity will occur, although it contributes to the possibility of that activity." (Greeno 1994)

As a result, the quality of design in the urban environment should be measured, validated, and related to its capacity in creating opportunities for complex activities that society realizes in absolute relation to the concept of urban habitability, as mentioned in this chapter. An urban environment should allow people to relate to each other, develop a sense of belonging and a sense of community. The infrastructure for active mobility must then work in this same direction, by designing environments and complete atmospheres of interaction and encounter.

According to Gehl (2006), the opportunities related to the mere fact of meeting, seeing and listening to other people include:

1) Contact in a modest level.
2) A possible starting point for contact in other levels.
3) A possibility to maintain already established contacts.
4) A source of information about the external social world.
5) A source of inspiration or an offer of a stimulating experience.

These are the reasons why it is important to consider the perception that people have of infrastructure. It is worthless to create infrastructure for active mobility if it does not foster human contact, or if instead of strengthening the sense of belonging and social empathy, it creates conflict among the different ways of mobility that should co-exist in the city. It is useless if it only moves people from one place to another just like the automobile does. The social component of active mobility, the urban habitability it promotes, should be considered from its beginnings and consciously taken care of, not as a random result, but as the main objective in the process of designing infrastructure.

PART III: THE IMPORTANCE OF DESIGNING URBAN INFRASTRUCTURE FOR ACTIVE MOBILITY AND ITS CONTRIBUTION TO THE HABITABILITY OF CITIES

The significant population growth in large cities has led to an accelerated urbanization – an urbanization that is sometimes well planned and in others, where the city grows directed by the real estate market, is based on the supply and demand of real estate.

According to Torres and Trujillo (2014), the social and economic dynamics of the cities, as well as the design based in favor of motorized mobility (private or public), "have contributed to the increase of a problem of their own such as transportation congestion."

Likewise, Perez-Gómez (2017) points out that planners have focused on creating urban networks that can accommodate traffic. This has led to the construction of more infrastructure for vehicles, resting importance from the urban infrastructure for active mobility.

Active mobility represents an alternative for growing cities; not only as a means of transport, but active mobility can contribute to the habitability of public spaces, to an adequate urban planning, and to the generation of urban surplus value (Martinez 2016). Not only does the construction of road infrastructure increase the value of real estate, but also building walkways, plazas and bikeways contribute to this growth, especially for local businesses by augmenting pedestrian traffic.

It is important to mention that the design of the infrastructure must be analyzed and considered from the point of view of the users. Both the needs of users of motorized vehicles in the design of roads (avenues, boulevards, streets) and of cyclists and pedestrians in the construction of sidewalks, bike lanes, walkways and open public spaces, serving as a link, must be considered as well as the interaction between them.

Rojas and Wong (2017) establish that "traveling by active modes of transport (i.e., walking and cycling) is affected by a series of factors, including the built environment and users' perceptions of these modes." The participation of citizens, who live in the urban environment and every day use more active mobility, helps to know their mobility needs. From this participation, it will be possible to provide the necessary tools to urban designers and government agents to generate public policies that encourage the use of non-motorized vehicles, as well as the construction of more and better infrastructure.

The needs of both types of users will lead to the creation of adequate urban infrastructure to stimulate the interaction of individuals with each other and within the same city. Additionally, this can promote different forms of mobility within the city.

The dream of every urban planner is to design a city where drivers of private vehicles, cyclists, pedestrians – and why not skate and skateboard users? – can interact and integrate with the environment. This is not only the work of the planner; other actors are also involved, for example, architects, civil engineers, government agencies, real estate developers, among others.

In addition to the urban design and to the aesthetic and functional aspects of cities, there is the concept of the Right to the City. This concept is mainly understood and developed by the legislative part of cities, by the rules and regulations that build the urban area. The regulations indicate the guidelines to be followed in order to provide infrastructure to the city.

The Right to the City plays a very important role. Thanks to this Right that citizens have, it makes them participants of their environment, whether as generators of public spaces from their professional activities or as participants with their way of life interacting with others. These give rise to the different atmospheres that are lived in the public spaces of interaction.

According to Padilla (2011), "the urban regulation must adjust to the social reality and go beyond those theoretical and limited objectives to give pre-eminence to the social functions of this activity." The planning, design and construction of public works is aimed at satisfying the needs of urban sprawl growth; it must respond to the current demands of the city, its habitability, and its citizens.

Despite the joint efforts to create cities, the design of urban infrastructure that responds to the Right to the City is still needed. Due to the lack of public policies that generate the adequate conditions for the design and construction of urban infrastructure, today we find divided public spaces, roads for car users, and spaces that serve for active mobility. Unfortunately, in some cities we continue to design and live in a world conceived mainly for motorists, downplaying the livability of cities (UN-HABITAT 2013).

It is not enough to design and build sidewalks or bike lanes for individual or collective mobility. As mentioned in the previous chapter, the design of infrastructure for active mobility is essential to generate a shift in the paradigm, considering the interactions between individuals in order to generate more livable cities.

The urban infrastructure must contribute to these interactions so necessary for the habitability of cities, improvement of the urban environment, and generation of active communication routes.

As mentioned by Jacobs (1961, quoted in Muxi et al. 2013), "cities are an immense laboratory of trial and error, failure and success, for construction and urban design." There is a general concern for having infrastructure just for automobiles; that is, roads that connect the opposite poles of the huge or small cities, whether these be narrow or wide roads. Sometimes it is road infrastructure constructed above the human scale of citizens – imposing bridges or isolated overpasses – which try to be an important link for vehicles. However, these separate those who try to make way for active mobility.

Efficient transport and good communication are not only difficult things to obtain; they are basic needs. It is imperative to deal with the design and construction of infrastructure for motorized transport and everything that promotes and facilitates active mobility.

It is not possible to separate vehicle and pedestrian traffic. We, however, have not been concerned with just generating adequate urban infrastructure for cities but rather, generating infrastructure that encourages social interaction. In addition, "while non-motorized trips may last longer than vehicular trips, for many developing city residents a higher time cost is preferable to a higher financial cost for transport" (Dimitriou and Gakenheimer 2001, quoted in Pojani and Stead 2015).

Unfortunately, we have concentrated on the construction of roads where the vehicle is the main actor. In recent years there have been initiatives with successes and failures where the gaze has been placed on the pedestrian and on collective transport. One such case is the Transmilenio Project in Bogotá, "which is the most visible physical change," providing time-saving, express public transport in the city (Bocarejo and Tafur 2013).

Copenhagen, Denmark, named the most livable city in 2013 by Monode magazine, has placed greater emphasis on walking and cycling (Centre for Livable Cities and Urban Land Institute 2014).

The cycling infrastructure that has been generated in some cities to encourage the use of bicycles has been inefficient due to the bad design of bikeways or their bad location. Moreover, in most governments there is not enough investment in this mean of public works.

Any infrastructure is transcendental for the development of the habitability and the functionality of the cities. This infrastructure that must be designed in a certain city will depend on the selection of the type of transport that the inhabitants use. This selection is based, mainly, on the economy and availability that users have for using certain means of transport, be it a private vehicle, bicycle or any other means.

Despite the importance of active mobility and its infrastructure, it is not regularly seen in the generation of urban policies in developing cities. To encourage active mobility, not only is it necessary to provide the adequate infrastructure (e.g., walkable environments, biking infrastructure and sidewalks) but also the willingness to change the behavior patterns of individual mobility (Haufe, et al. 2016).

In some cities, residents still do not use the bicycle as a means of transportation. In scattered cities it is more difficult to move in non-motorized vehicles or to walk, due to the great distances that have to be covered. Although in some compact cities distances are shorter, users feel unsafe on the streets due to poor infrastructure.

As already mentioned in previous paragraphs, it is also important to instigate public policies and legal instruments to promote the inclusion of active mobility in urban development plans. If all these factors are promoted, habitability can take place in cities. "Such knowledge can help policymakers and authorities to evaluate the level-of-service of current active-mobility network" (Rojas and Wong 2017).

Returning to the concept of habitability and its importance, it is essential to understand current and potential users as well as their attitudes and perceptions towards active mobility. According to Rojas and Wong (2017), this understanding can lead urban planners and developers, engineers, architects and authorities to develop adequate infrastructure schemes to integrate motorized and non-motorized mobility.

Conclusion

Our contribution has presented the importance of active mobility, the adequate construction of its infrastructure and the essentialness of achieving the integration of active of mobility with the motorized type, thus ensuring that individuals are integrated and live the city from a different perspective. The foregoing establishes that habitability is the main actor in the city.

Jacobs (1961, quoted in Muxi 2013) embraces the idea that citizens must feel and live the city. How do people interact and feel the city? It is a subject that has not yet had the necessary strength for individuals to live in community, to live a sense of belonging, and to empathize with each other.

The design and construction of infrastructure for active mobility will lead cities to have a better urban growth and to have better combinations of communication routes for motorized and non-motorized transport. This will contribute significantly to the habitability of cities.

It is imperative to promote integration among all the actors that create the city, with the aim of taking into account the importance of active mobility. With active mobility, costs and pollution are reduced; the health of the citizens is improved; and also habitability, in having public spaces, individuals can interact and have a better life.

It may be noted that mobility is not only a source of identity, as some may think, where having a private vehicle provides a sort of membership to a middle, medium or high socioeconomic level. Active mobility, in addition to giving identity to the city, can generate great benefits in the health of citizens and have positive effects on the environment. Mobility represents a starting point for the development of sustainable, efficient, and dynamic cities.

Finally, active mobility can grow and improve if users of this type of mobility manifest their need to have the necessary infrastructure for this type of transport. Also, it is possible if authorities promote the use of public transport and decrease the use of private vehicles. This way, we will achieve integral, livable cities with better public spaces to interact.

REFERENCES

Bocarejo, Juan Pablo., and Tafur, Luis Eduardo., 2013. *Urban land use transformation driven by an Innovative Transportation Project, Bogotá, Colombia*. Nairobi: Case study prepared for Global Report on Human Settlements. Accessed January 20, 2018. https://unhabitat.org/wp-content/uploads/2013/06/GRHS.2013.Case_.Study_.Bogota.Colombia.pdf.

Centre for Liveable Cities and Urban Land Institute. 2014. *Creating healthy places through active mobility*. Enviro Wove. Accessed January 21, 2018. http://clc.gov.sg/documents/books/active_mobility/index.html.

Church, Edward., Carolina, Miranda., Nicola, Szibbo., Rebecca, Elliott., Galen, Cranz., 2012. *Cities and People Project: A White Paper on Human Interaction with the Built Environment*. Institute for Environmental Entrepreneurship Berkeley, California, U.S.A.

Erner, Guillaume., 2013. *Sociología de las tendencias*. [*Sociology of trends*] Barcelona: Gustavo Gili.

Flores-Gutiérrez, Avatar., 2016. *Fenómeno arquitectónico, proceso de diseño y complejidad humana. Propuesta de re-conceptualización.* [*Architectural phenomenon, design process and human complexity. Proposal for re-conceptualization*] PhD diss., Universidad Nacional Autónoma de México (UNAM), Ciudad de México.

Gehl, Jan., 2006. *La humanización del espacio urbano.* [*The humanization of urban space*] Barcelona: Editorial Reverté.

Gifford, Robert., 2007. *Environmental Psychology. Principles and Practice* (Fourth edition) Canada: Optimal Books.

Greeno, James G., 1994. *Gibson's Affordances. Psychological Review*, Vol. 101, No. 2, pp. 336 – 342.

Haufe, Nadine., Milloning, Alexandra., and Markvica, Karin., 2016. *Developing encouragement strategies for active mobility. Transportation Research Procedia* 19. 49-57. Accessed January 21, 2018. https://ac.els-cdn.com/S2352146516308535/1-s2.0-S235214651

6308535-main.pdf?_tid=71a2e30e-04b6-11e8-9fab-00000aacb361&acdnat=1517204407_15cab10102f65ee8b8e75c6e724e9d15.

Hess, F., Salze, P., Weber, C., Feuillet, T., Charreire, H., Menai, M., et al. 2017. Active Mobility and Environment: A Pilot Qualitative Study for the Design of a New Questionnaire. *PLoS ONE* 12(1): e0168986. https://doi.org/10.1371/journal.pone.0168986.

López, Riquelme., Germán, Octavio., 2018. Biofilia: Cimientos para un urbanismo evolucionista. [Biophilia: Foundations for an evolutionary urbanism] *Paper presented at the "1era jornada de arquitectura y cognición."* Centro de Investigación en Ciencias Cognitivas (CINCCO): Universidad Autónoma del Estado de Morelos. January 24.

Martínez, Catalina., 2016. Movilidad es un reto fundamental para las ciudades en crecimiento. [Mobility is a fundamental challenge for growing cities.] *Inmobiliare Magazine.* Desarrollo urbano. Accessed January 20, 2018. https://inmobiliare.com/movilidad-es-un-reto-fundamental-para-las-ciudades-en-crecimiento/.

Muxí, Zaída., Valdivia, Blanca,, and Delgado, Manuel., 2013. *Muerte y vida de las grandes ciudades. [Death and life of the big cities.]* Capitán Swing Libreos. Third edition. España.

Padilla, José., 2011. Los servicios públicos municipales la obra pública. [The municipal public services public works.] In *Derecho Urbanístico,* edited by Jorge Fernández and Juan Rivera, 82. México. Universidad Autónoma Nacional Autónoma de México.

Pérez-Gómez, Alberto., 2017. *Attunement. Architectural Meaning after the Crisis of Modern Science.* Boston: MIT Press.

Pojani, Dorina., and Stead, Dominic., 2015. *Sustainable urban transport in the developing world: Beyond megacities. Sustainability* 7,7784-7805. Accessed January 26, 2018. doi:10.3390/su7067784.

Rojas, María Cecilia., and Wong, Yiij Diew., 2017. Attitudes toward active mobility in Singapore: A qualitative study. *Case Studies on Transport Policy,* 5, 4, 662-670.

Torres, M., and Trujillo, J., 2014. La ciudad y su dinámica. [The city and its dynamics.] Orinoquia. Universidad de Llanos. *Villavicencio, Meta. Colombia.* Vol. 18. No. 2. p.9.

UN-HABITAT. 2013. *State of the World´s Cities 2012/13-prosperity of cities, Routledge.* Accessed January 25, 2018.

Vasconcellos, Eduardo Alcántara., 2011. *Desarrollo urbano y movilidad en América Latina* [*Urban development and mobility in Latin America*]. Bogotá: Corporacion Andina de Fomento.

Zumthor, Peter., 2010. *Atmósferas. Entornos arquitectónicos – Las cosas a mi alrededor* [*Atmospheres Architectural environments - Things around me*]. Barcelona: Gustavo Gili. Accessed July 2014.

In: Transportation Infrastructure
Editor: S. Antonio Obregón Biosca
ISBN: 978-1-53614-059-0
© 2018 Nova Science Publishers, Inc.

Chapter 3

ADVANCED ASSESSMENT AND STRATEGIC MANAGEMENT OF PAVED ROAD NETWORKS: THE CASE OF COSTA RICA'S NATIONAL ROAD NETWORK

José D. Rodríguez-Morera[*]
and Luis G. Loría-Salazar[†]*, PhD*
LanammeUCR, Universidad de Costa Rica, San José, Costa Rica

ABSTRACT

This chapter presents transportation asset management principles and philosophy, focused on pavement management systems. Advanced methodologies for assessing paved road networks using condition indices like International Roughness Index (IRI), deflections obtained from the falling weight deflectometer and skid resistance, as well as management practices to be developed from the results of the assessments are described.

---

[*] Corresponding Author Email: jose.rodriguezmorera@ucr.ac.cr.
[†] Corresponding Author Email: luis.loriasalazar@ucr.ac.cr.

The assessment of the National Paved Road Network of Costa Rica consisting of more than 5000 km and carried out bienally by the National Laboratory of Materials and Structural Models (LanammeUCR) of the University of Costa Rica, is included as a case study. This evaluation, performed seven times since 2004, shows clearly the potential of pavement management and how to achieve efficient and effective investments, once the implementation challenges have been overcome.

The inputs produced by LanammeUCR are high quality data to improve the decision-making processes, such as maintenance strategies, budget allocation and accountability, all with a strategic level approach that allows an accurate view of the evolution of the national paved road network condition in the long term.

Keywords: pavement, asset, management, road network, assessment

INTRODUCTION

Road Network Management with budget constraints is a common issue that Transportation Agencies face. To apply sound infrastructure policies and achieve performance goals and quality level of service, institutions in charge should turn their eyes to the asset management philosophy.

Transportation Asset Management operates data-based decision-making processes, considering preventive maintenance and looking to obtain the best condition possible on assets with the available funds. These principles are applied in Pavement Management, which is possibly the largest asset in a transportation system, using different approaches according to the level of management: network level or project level.

Data bases are considered the heart of asset management systems and they consist of information about the asset, in this case pavements. In this sense, data collection and the quality of data obtained are fundamental to produce high quality analysis and forecasts. Paved Road Network assessments are the way a Pavement Management Systems creates its data base. To perform this assessments process can vary from visual inspection to advanced techniques using high tech equipment.

To better visualize Pavement Management process, besides presenting an asset and pavement management theoretical framework, the case of the oversight model created by Law 8114 in Costa Rica, that grants the National Laboratory of Materials and Structural Models (LanammeUCR), is explained herein as a Pavement Management ally in this Latin American country.

TRANSPORTATION ASSET MANAGEMENT

While budgets of transportation agencies are limited, maintenance needs of road networks are almost infinite. This reality caused many transportation agencies turned their look towards private sector business practices. That is how the philosophy behind transportation asset management was born. According to the Asset Management Subcommittee of the American Association of State Highway and Transportation Officials (AASHTO), Transportation Asset Management can be defined as:

> "A strategic and systematic process of operation, maintenance, improvement and expansion of assets; which runs through the lifecycle of the assets. It is based on business practices and engineering for the allocation of resources, with the aim of making the best decision based on quality information and well-defined objectives".

Asset management philosophy "embrace all the processes, tools, data and policies necessary to achieve the goal of effectively managing assets. In general, the sequence of asset management begins with the identification of goals and policies of a road administration and the available budget. This starting position, the sequence proceeds through data collection, performance monitoring, analysis of maintenance options and program optimization, through to project selection and implementation" (OECD, 2001:9).

The basis of transportation asset management is proactive maintenance. Therefore, the aim is to eliminate the "worst first" approach that consists in letting the condition of assets in "good condition" decline, while spending money in assets in worse condition at a higher cost

(FHWA, 1999). Prioritization criteria are a key to implement successfully a transportation asset management business approach.

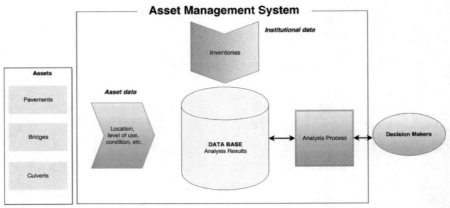

Source: OECD (2001).

Figure 1. Typical Flown of Data into and out of a generic road management system.

According to AASHTO (2011), to implement the asset management philosophy in an agency, a Transportation Asset Management System (TAMS) is required. TAMS are transversally shaped by three main axes: organization, human resources and technology. In addition, each of these axes comprises a structure of three levels called strategic level (executive), tactical level and operational level (project).

At the strategic level (with a planning time frame of 20 years), decisions are made regarding policies, performance goals and strategies, having an integral approach on road assets (bridges, walls, culverts, pavements, among others). Which routes will be prioritized? Which routes should be expanded? What level of investment will the Transportation Agency maintain in the next 20 years? What standard should the roads have according to their hierarchy? These questions must be answered at the strategic level of a road agency.

The tactical level, with a planning time frame of five years, establish the link between the strategic level and the operational level. It is the level where the translation of the policies into projects is carried out to contribute to achieve the goals defined for the country or region. At this level, activities such as feasibility studies, expropriation procedures, preparation of bidding and hiring processes are carried out.

Last, the operational level is the level of execution and therefore its planning time frame can be from 1 to 3 years, depending on the complexity of the project but mostly it consists of annual plans. At this level, projects must arrive ready so that the executors do not suffer a setback due to the absence of studies, contracts with a partial scope, pending expropriations or relocation of public services infrastructures not included before. This level is also fundamental in the feedback for the planning process, since the condition of the roads once they have been treated must be reported.

On the other hand, within the axes that support this TAMS structure, the most important is the organizational one. An institution can count on trained professionals and technology tools, but if its processes, communication strategies, roles and responsibilities are not clearly defined, the organization will be doomed to high rates of reprocessing, duplicities and low efficiency.

Regarding the human resource component, professional development, continuous training and a sound incentive regime that aligns the organization around proactive management and obtaining results are essential for the sustainability of TAMS.

About technology, TAMS development and implementation guidelines published by AASHTO, indicate that it is the easiest component to implement. In fact, software tools are the last step of an implementation process in accordance with these guidelines, and not the other way around, as usually happens in the Latin American context. Software should be adaptable to organizations and not conversely. In general, databases conform the heart of TAMS and they are an important part of the technology domain.

Summarizing the core processes mentioned above in terms of data from the principal axes that formed a TAMS, on Figure 1 is shown the typical flow this data has in and out of a generic road asset management system. Every level of management in a TAMS produce information and it must be well collected to create a solid data-based decision-making process.

PAVEMENT MANAGEMENT SYSTEMS

A Pavement Management System (PMS) is an individual asset management system within a global system of Transportation Asset Management. According to Solminihac (1998), a Pavement Management System is the set of operations aimed to conserve for a period the conditions of safety, comfort and structural capacity suitable for circulation, enduring climatic and environmental conditions in the area in which the road is located. All of that, minimizing monetary, social and ecological costs.

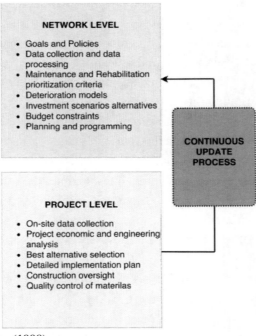

Source: Solminihac (1998).

Figure 2. Main activities in each levels of management in a PMS.

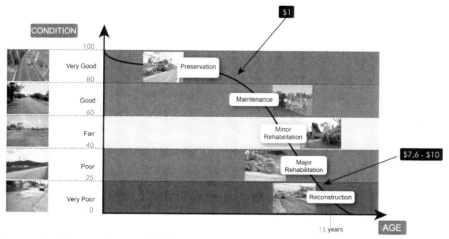

Source: Rodríguez-Morera (2012).

Figure 3. Cost variation according to pavement condition estimated for Costa Rica.

For the Transportation Association of Canada (TAC, 1997), the purpose of pavement management is to achieve the best possible condition in accordance with the public funds available. This is accomplished by comparing investment alternatives both at the network level and at the project level, which are the two levels that form a PMS.

Management at the Network level allows to determine the needs in a set of roads. At this level, a priority and organized program of rehabilitation, maintenance or construction of new paved roads considering budget constraints is developed. On the other hand, management at the project level clearly defines the requirements of a specific project. The Figure 2 shows the differences between the activities in each one of the levels in a PMS.

Pavement Management Systems have changed the approach from reactive to preventive attention of paved road networks. This means that the practice of treating "worst first" get abandoned. As a result, users perceive an improvement in mobility, comfort and road safety (FHWA, 2007). Experiences with pavement conservation in several States in the U.S. have shown this success: every dollar that is currently spent on pavement conservation could save up to six dollars in the future (FHWA,

2007). This ratio was calculated for the case of Costa Rica (Rodríguez-Morera, 2012) and the increase of cost by delaying interventions during pavement lifespan is bigger as show in Figure 3.

The Three Principles of Pavement Management

Pavement maintenance strategies are not suitable for pavements that require major rehabilitation or reconstruction. In these cases where the life span of the pavement cannot be prolonged further or where deterioration is severe, conservation practices will be ineffective in technical and economic terms. In addition, user comfort will be compromised because the surface condition in these cases will be poorer.

At this point, we must remember the three fundamental principles of Pavement Management: apply the right technique, in the right place at the right time. This means that maintenance techniques required by pavements must be chosen according to its condition, in the section of the road that needs to be attended (based on sound prioritization criteria) at the right time of the life span, where it does not imply greater investments due to an advanced state of deterioration.

No maintenance techniques will prevent pavement deterioration forever, however, conservation strategies and techniques can significantly slow down deterioration (FHWA, 2007). To extend the life span of pavements it will be necessary to guide the management practices under these three principles.

Data Bases and Inventories of Paved Road Networks

The heart of Pavement Management Systems (PMS) are databases that contain the inventories of paved road networks. PMS require an inventory of the road network or project to be analyzed. This should contain the permanent characteristics of the pavement as location, structure and geometry (Solminihac, 1998).

Transportation agencies should "standardize data collection and establish a common reference method to identify sections that allow information to be managed efficiently" (Solminihac, 1998). Once the sections or units under which the road network will be managed are created, periodic evaluations should be carried out to update the condition of the pavement in each section as well as document investments, results and any other relevant information.

Standardization is fundamental because it allows to integrate the information. If an Agency have to manage a very extended road network possibly, the road network will be divided into regions. The inventory of every region needs to be integrated and that can only be done in an efficient fashion if data collection criteria is the same and the file formats are also the same.

Road network sectioning can be done based on several criteria, for example geographic references, surface and pavement structure types and even assets condition. Once the sectioning is defined, it is important to develop a strategy to collect in the future all road assets data further than the pavement. This is where it is important to consider data base design and file formats to manage it all in a comprehensive way.

Evaluations of paved road networks are the process to keep the heart of the system (databases) in shape, generating constant flows of information that updates knowledge about the network. Like other processes, evaluations of paved road networks can be done at the network or project level, depending on the need of the transport agency or the mandate received by the executing agency.

Pavement Management Systems are based on the results obtained from objective performance measure parameters to determine the condition of the pavement. However, for the information generated in the evaluations to be valid and of course, feasible to be obtained, it is essential to define whether the analysis will be at network level or at project level.

As shown in Figure 4, if pavement management is done at project level, the detail of information and the complexity of the models must be greater because the studies are more specific; moreover, when planning at a strategic or network level, it is not feasible to use detailed information or complex models because it will generate an excessive amount of results that do not impact on the quality or the objectives of the analysis.

Authors such as Flintsch (2007) have described five levels of information in road management that correlates the degree of sophistication and detail required for each decision-making and data processing stage. As presented in Figure 5, level 1 is the one that requires more detailed information, more frequently used for research and lab projects. In Level 2, typical information for engineering analysis and project level decisions will be found. Level 3 is a simpler level of detail, that refers to generally two or three attributes that can be used at the network level or in the collection of more general data, this includes friction deterioration. Level 4 is a summary of key attributes that are used in planning and executive reports for senior management staff. At level 4, structural and functional performance are evaluated. Finally, Level 5 represents executive higher-level and less detailed data, such as key indicators of transportation assets performance.

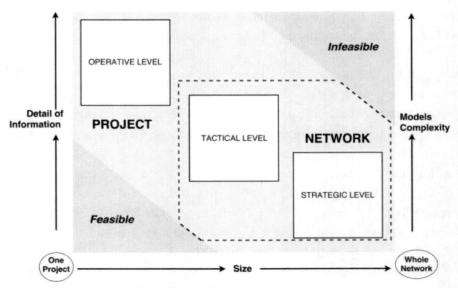

Source: Adaptation from Solminihac (1998).

Figure 4. Detail of information and complexity for a pavement management system.

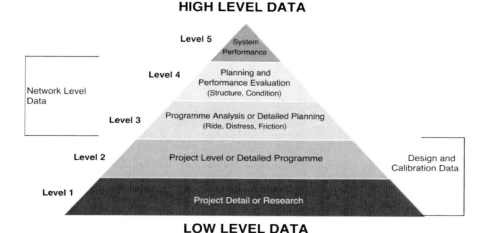

Source: Flintsch (2007).

Figure 5. Levels of information quality for road management.

ADVANCED PAVEMENT EVALUATION

The evaluations of pavements allow to determine the functional, structural or superficial condition of roadways. One of the most important aspects is to define the use of the information that is going to be collected. The detail needed in the data will be defined according to the level of management.

In addition to knowing the pavement condition, evaluations allow investment planning on road maintenance, leading to an adequate road network management and generate the possibility to road users, authorities and stakeholders to demand efficiency in budget allocation.

There is a wide range of technologies that allow the control of road network characteristics. However, it is necessary to properly select the equipment, climate conditions and the way in which the data collected will be managed. Usually, roughness, texture, skid resistance, mechanical and structural properties, superficial pavement distresses and road geometry are evaluated. Therefore, the information generated by these evaluations is important because it allows the determination of pavement condition indexes that facilitate the understanding and accountability about the paved road network conditions.

Within the main characteristics controlled, it is found the pavement structural capacity. This characteristic is obtained through the deflections suffered by the pavement in response to an applied load that represents the weight of the vehicles. The results are related to damages such as cracking and permanent deformation. One of the equipment used for the purpose is the Falling Weight Deflectometer (FWD), which is a high-tech equipment that measures the subsidence or instantaneous deflections experienced by the pavement at one point, in response to the blow of a falling weight that produces a force of 40 kN (see Figure 6).

Surface roughness is a determining factor in a pavement evaluation. This characteristic is the sum of the irregularities of the pavement surface per length unit, which is perceived by the user as riding comfort (Solminihac, 1998). It is related to the level of service, but it is also an indicator of operating system deficiencies, security and travel speed.

Source: LanammeUCR (2015).

Figure 6. Falling Weight Deflectometer used by LanammeUCR.

The most used and internationally accepted parameter to measure surface roughness is the International Roughness Index (IRI), which is a standardized measure of roughness that uses a mathematical model known as Reference Quarter Car Simulation (RQCS) for its determination. In this model, vertical movements of the suspension are accumulated and divided into the traveled distance, the units of the results are "m/km" or "inch/mile".

The quarter-car model used in the IRI algorithm is a model of one corner (a quarter) of a car. The model "includes one tire represented with a vertical spring, the mass of the axle supported by the tire, a suspension spring and damper, and the mass of the body supported by the suspension for that tire" (Sayers y Karamihas, 1998: 50).

At road network level or project level evaluation, a vehicle equipped with a laser profiler (as shown on Figure 7) is used. It determines in real time the IRI value in 100 m length sections. It is expected that a newly constructed road will present a minimum IRI value and as soon as its use increases, the loads of vehicles will increase roughness and therefore the value of IRI also (Barrantes et al., 2008).

Source: Sanabria, J., Barrantes, R. and Loría-Salazar, L. (2015).

Figure 7. Laser Profilometer to measure IRI used by LanammeUCR.

Source: Sanabria, J., Barrantes, R. and Loría-Salazar, L. (2014).

Figure 8. Grip tester used in LanammeUCR.

Another relevant parameter in the condition of the pavement is the friction of the surface, measured using a grip tester (See Figure 8). This refers to the adherence between the tire of the vehicle and the pavement surface. This factor is important for the safety of road user's due to the skid resistance of the vehicle and the reduction to the risk of loss of control. Maintaining a minimum acceptable surface friction value is vital to maintain the normal service and safety conditions of a road (LanammeUCR, 2015).

The skid resistance depends on the micro texture and the macrotexture of the pavement. The micro texture is the surface of the aggregate that generates friction force when it makes contact with the tires. In wet conditions, the micro texture penetrates the thin layer of water between the tire and the road, which is why the skid resistance dominates at low speeds (less than 70 km/h). However, the macrotexture, which is related to the type of aggregate exposed in the mixture, is what allows water drainage and prevents hydroplaning at high speeds.

Source: Sanabria, J., Barrantes, R. and Loría-Salazar, L. (2014).

Figure 9. Geo3D equipment to collect surface distresses visually.

Finally, paved road networks inventories are fed with several types of information, not only those coming from pavement structures. Other kind of evaluation such as visual assessment identifying surface distresses, marking deficiencies and even road safety issues can be carry out.

For this objective, vehicles equipped with 360° view like the one shown in Figure 9 are used to assess, both network and project level. This data will complement pavement sections information.

IMPLEMENTATION OF TRANSPORTATION ASSET MANAGEMENT SYSTEMS

The implementation of TAMS, as in the case of a Pavement Management System, represents an important change in the practices of a traditional organization. To face this challenge, it is recommended to develop a strategy that contains actions on each of the areas where the changes will be implemented. For this strategy to be as effective as possible, it is recommended to start with a self-assessment that allows to know where the transportation agency or road authority is, on a scale of asset management maturity.

According to the "Transportation Asset Management Guide: a focus on implementation" (AASHTO, 2011), it is necessary to develop several steps to successfully implement the system in a transportation agency or ministry. These steps are grouped into four major areas: strategic direction, alignment of the organization, long-term planning, and as a last step, processes of technological tools, i.e., software and IT systems.

Generally, there is resistance when it is decided to implement a Transportation Asset Management System because the changes that must be made produce great challenges and seems difficult at the organizational level. Organizational changes represent bigger challenges than the technical part, therefore, strategies that can bring change more effectively must be discussed, establishing trust and confidence among managers and teams. Some means to achieve this are workshops, group sessions, follow-up and monitoring for feedback from collaborators (FHWA, 2007).

Furthermore, "for the senior executive, whose term at office may be of limited duration, asset management implementation may be best conceived as an organizational change project, with a star point and objective on a two to four-year time scale". Thus, political leadership can shift from threat to support, champion the change to TAMS (AASHTO, 2011: 1-4).

Barriers to Pavement Management Systems

Barriers may arise against the implementation of a TAMS and a Pavement Management System is not the exception. As it has been said, it is possible that technical difficulties are the easiest to overcome, especially when we talk of a change in the way a transportation agency do their business. Hence, below some barriers are described to be considered in the design of strategies to overcome them.

- Technical barriers: within the technical difficulties there are those related to the quality of the information of the assets of the road network, for example pavement, bridges, culverts, signs. Besides that, data collection methods and equipment, and databases design could be other technical issues.
- It must be remembered that one principle in infrastructure management is "keep it simple before fancy". This means that we must start with the available resources and then design both, databases and the software systems, in a way they can be expanded and get more sophisticated as the maturity of the organization increases.
- Financial barriers: budget will always be a restriction because you never have all money you need. This instead of being an obstacle, must be the motivation to implement asset management philosophy because precisely, it seeks to optimize budgets by making better decisions.
- The decision to dedicate part of the budget to the implementation of a Pavement Management System, must be based on the benefits that will be obtained. For this, it is desirable to perform long-term comparison analysis of scenarios that show the differences between the results from the actual practices and those that could be gained by changing to an asset management strategy.
- Political barriers: convince elected officials is a feared challenge from the technical staff. To overcome this barrier, we must bet on a strategy to find allies. It is essential to have "champions" of TAMS philosophy that have sufficient leadership and communication skills to convince elected officials. Fortunately, there are numerous successful stories that can be used as an example of good practices and as a demonstration of where the transport agency can be.
- Cultural barriers: organizational change is the key to advance in the level of maturity of a transport agency in terms of management of transportation assets. There are different methodologies and approaches to implement the change, but a key will always be the

information. Excess information will not hurt as the lack of information will, so it is necessary that information on the processes of change be always available and designed according to the target audience, from project staff to legislators and stakeholders.
- Legal barriers: in some institutions there may be legal restrictions on some practices of TAMS. In this regard, an extensive study was published by AASHTO, presenting the state legislation in the United States on this subject, which shows the tools available to the Departments of Transportation. These tools range from financing and accountability acts to the definition of pavement performance indicators.

Best advice to give in regard of these barriers is to remember there is no single recipe on the design of TAMS framework. Therefore, each country and each institution must agree on the one that best adapts to its context and allows to obtain the best results from the implementation asset management philosophy.

Table 1. Benefits of Transportation Asset Management Systems

Economic	Performance	Accountability	Communication	Staff
• Lower long-term costs for infrastructure preservation. • Build, preserve, and operate facilities more cost effectively with improved performance • Improved budget process.	• Improved performance and service to customers. • Improved cost-effectiveness and use of available resources. • A focus on performance and outcomes • Better asset inventory, condition and level of use. Road network performance.	• Improved credibility and accountability for decision making processes. • Enhance institutional credibility.	• Communicate plan for growth • Maximize the benefits of Infrastructure • Provides the desired levels of service. • Long term view • Better communication (both internal and external to the administration).	• Clear relationships, transparency and accountability • Staff development.

Source: AASHTO (2011) & OECD (2001).

BENEFITS OF PAVEMENT MANAGEMENT SYSTEMS

It is true that the implementation of a TAMS has its cost, as the economic investments necessary for organizational changes, training and technology components. However, internationally, the benefits of this management philosophy have been demonstrated. Table 1 shows some of the benefits obtained after implementing TAMS in a transportation agency.

One of the key functions of asset management is to measure its own benefits in terms of performance, a capability that otherwise would not exist in most transportation agencies (AASHTO, 2011:1-13). Consider and publicize these benefits at the implementation of TAMS will help to gain stakeholders support and approvals on funding.

THE CASE OF LANAMMEUCR: BIENNIALLY ASSESSMENT OF THE NATIONAL PAVED ROAD NETWORK IN COSTA RICA

The case herein presented is the experience generated by the Transportation Infrastructure Program (PITRA) of the National Laboratory of Materials and Structural Models of the University of Costa Rica (LanammeUCR). This Laboratory is an academic research founded entity in the 1950s and linked to the School of Civil Engineering of the University of Costa Rica. The LanammeUCR is a National Laboratory specialized in applied research and training in the field of civil infrastructure, structural engineering, road materials transportation and road engineering.

LanammeUCR has as a legal framework created by the Law 8114 of Tax Simplification and Efficiency, which grant the University of Costa Rica (UCR), the oversight faculty for the national road network and the financing to execute it (1% of the fuel tax). Through an external and independent entity, in this case, the University of Costa Rica, legislators determined the way to guarantee to the country the quality of Costa Rican national road network and the maximum efficiency of public investments on construction and maintenance road works.

As the University of Costa Rica is an autonomous entity according to a constitutional precept, the LanammeUCR plays a key role in strengthening accountability, with plenty of powers to perform their activities with functional independence, under the principles of objectivity.

This is how this road oversight model was developed, generating processes and products that became a fundamental ally for the Administration in the field of Pavement Management. For example, through this oversight model, more than 30000 km in the National Paved Road Network of Costa Rica have been assessed, producing information about its condition and its evolution.

The evaluation of the paved road network that is carried out biennially according to Law 8114, has been performed seven times. In addition, more than 1100 road audit findings have been reported to the Administration. Each report that LanammeUCR releases is sent, in accordance with the Law 8114, to the Ministry of the Presidency, Ministry of Public Works and Transportation, Road Authority (CONAVI), Legislative Assembly and to the Office of the Ombudsman. Therefore, accountability is a strong component within this road audit model.

With this background, results and evolution of the LanammeUCR paved road network assessment are presented to demonstrate its importance within a pavement management approach to improve paved roads quality in Costa Rica.

Results of the Paved Road Network Biennially Assessment in Costa Rica

The results presented by LanammeUCR in their reports serve as a management tool for the Administration (Ministry of Public Works and Transportation and the Road Authority). In addition, it is an input for planning and also helps the Road Authority to resolve disputes with contractors.

The parameters chosen to carry out the evaluation are directly related to the roads lifespan and to the operational costs of the vehicles. The Falling Weight Deflectometer (FWD) measures the deflections obtained by applying to the pavement a force that simulates traffic loads, these measurements are related to the bearing capacity and to the remaining lifespan.

As part of the research carried out in 2008 in the LanammeUCR, ranges of deflectometry were developed based on the Average Daily Traffic (ADT) that each route presents. These ranges are intended to represent more accurately the actual conditions of use of national routes, so they were used in the evaluation campaign to classify the results.

The Laser Profilometer measures the surface roughness obtained in a specific road section. This measure is associated both with the comfort felt by road users traveling through this section, and to the operational costs of the vehicles.

The first evaluation of the paved national road network was carried out in 2004, although it was in 2002 when a partial evaluation was first performed to calibrate the equipment. In 2004, approximately 4,000 kilometers (4,081.3 km with the laser profilometer and 3,776.8 km with the FWD) were surveyed and evaluated. The results of this evaluation campaign showed that in terms of surface condition, approximately one third of the road network had good conditions, another third had regular conditions and the remaining third had poor conditions; while the results with the FWD indicated that only 13.6% of the evaluated length had low deflections (desirable condition), 22.1% moderate deflections and the remaining 64.3% high and very high deflections (Sanabria, J., Barrantes, R. and Loría-Salazar, L., 2015). This is how the first evaluation campaign showed the advanced level of deterioration of the national paved road network in 2004 and the ineffectiveness of the measures implemented to maintain and recover the road network.

The second assessment campaign of the road network was carried out in 2006, the results showed that, according to the data measured with the laser profilometer, the surface condition of the national paved road network was such that approximately one third of the length was found in each of the superficial regularity ranges defined as: good, regular and poor (Sanabria, J., Barrantes, R. and Loría-Salazar, L., 2006).

Also, according to the evaluation with the FWD, less than 12% of the road network had low deflections (desirable condition), 22% had moderate deflections, and 64% high deflections. When comparing the results obtained in that evaluation campaign with those of the 2004 evaluation campaign, it was determined that the state of the road network had not experienced an appreciable improvement during the last two years, on the contrary, a tendency toward deterioration was detected. These results indicated that pavement management carried out during the two previous years on the road network had not been effective in the recovery of road assets, so it was necessary to make changes in the way in which the resources destined to the road infrastructure of the road were managed and executed.

As happened with the campaign of 2006, the evaluation carried out in 2008 was implemented using Global Positioning System (GPS), to improve the information related to the location of the measurements. Additionally, the GPS facilitated the storage of data in the Geographic Information Systems (GIS), which allowed a more agile handling and analysis of information thanks to the collaboration of the Transportation Sector Planning Office at the Ministry of Public Works and Transportation. The data obtained in this campaign had the basic information of the control sections, which simplified the communication of the information to the public.

It was noted that two thirds of the road network had deflections within the low or adequate range, which was established according to the average daily traffic (ADT), which allows to give a more realistic idea about the state of the routes according to the intensity of traffic. This evaluation revealed that 231 km improved in this parameter between 2006 and 2008. However, about one fifth of the road network (828 km) had high and very high deflections, which revealed a possible weak pavement, construction deficiency of the route, or an advanced state of deterioration. The IRI surface regularity evaluation showed that only one fifth of the road network had an acceptable IRI value; one third a moderate value; and a little more than half the high values in this parameter which was performed only during the day, covered 83.7% of national roads.

A serious deficiency in road safety was noted given that only a third of the 3,878 km evaluated presented a pavement marking from fair to good; while about half of the evaluated routes showed no pavement marking. It meant that two thirds of the road network (2,564 km) had a deficient marking in 2008, which reduces the safety conditions for road users (LanammeUCR, 2008).

Regarding the evaluation of the paved road network in 2010, for the IRI, 444.8 km more than in 2008 were evaluated. This difference is due to the increase of the paved routes and the inclusion of sections of municipal routes within national routes. In this case, the national road network showed a significant improvement in terms of skid resistance, in addition, surface distresses were detected.

For the 2015 evaluation, the results showed that 90.22% of the evaluated road network was in good condition according to the deflection parameter FWD, the deflections that exceeded the moderate state were 5.47% of the Road Network. Regarding the results of IRI, 37.58 km represent 4.51% of the Road Network with values of good surface roughness, while the category of fair condition covers 1,761.85 km for a 33.44%. The remaining 62.05% of the Road Network was found in poor and very poor conditions of surface roughness.

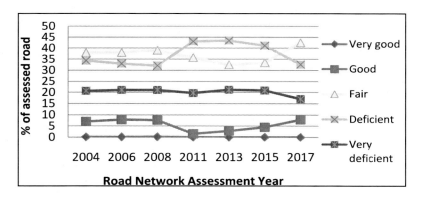

Figure 10. Evolution of IRI of Costa Rica´s National Paved Road Network.

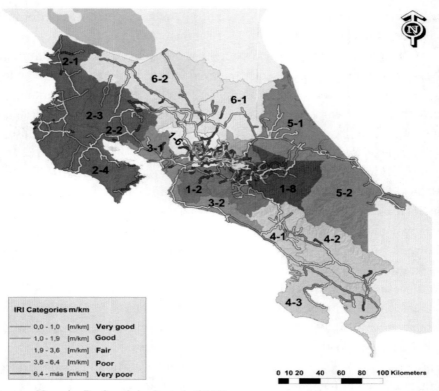

Source: Naranjo, R., Sanabria, J. et al. (2017).

Figure 11. Map of IRI condition of Costa Rica´s National Paved Road Network in 2017.

The most recent campaign of 2017, found that 84.5% of the road network has good structural capacity, the moderate condition is presented in 6.11% of the network. According to 2017 report, Maintenance actions must be designed to improve or maintain their condition to prevent deterioration. Besides, superficial roughness, 49.65% have poor and very poor roughness standards, however, they present an improvement compared to the evaluation made in 2013.

As shown in Figure 10, due to the assessment performed by LanammeUCR, Costa Rica can track the variation of the IRI in almost its whole paved national road network (about 5500 km) since 2004. This data base is a highly valuable information to pavement management that can be used in several ways, depending on the stakeholders. Because of the nature of LanammeUCR, this is information is public and can be obtained through a web site by any person, including consultants, students, media, elected officials and public in general.

Using GPS tools, all the information of the pavement assessments is presented in maps. As it was said, this kind of presentation is very useful at network level in pavement management because it communicates with the enough detail what stakeholders want to know. Color scales are used to show the roughness condition in the national paved road network in the Figure 11. Other maps presenting FWD, skid resistance and pavement marking condition are included in LanammeUCR reports.

As it could be observed, the evaluation process of the national network has been refined over the years. Based on the campaigns carried out, and those that will be carried out in the future, the deterioration or recovery process experienced by the road infrastructure can be monitored. In addition, these can be used as an input to prioritize routes or areas for their prompt intervention, so it influences decision-making processes at technical, economic and political level.

LanammeUCR Paved Road Network Assessment as a Pavement Management Ally: Maintenance Strategies and Budgetary Accountability

Besides the oversight task that LanammeUCR performs, a proposal to define intervention strategies at the network level have been developed in recent reports. These strategies developed by the Road Evaluation Unit in PITRA (LanammeUCR) come from the combination of the calculated IRI values and the deflection values obtained with the FWD, that reveal the structural capacity of the roads. The combination of these two parameters defines a series of "Q" (quality categorization) for national paved routes. These Q´s (See Figure 12) establish different levels of deterioration for the roads, allowing to define if a section is eligible to be intervened through different activities such as preservation, maintenance, rehabilitation or reconstruction (See Figure 13).

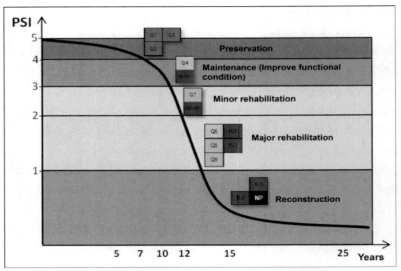

Source: LanammeUCR (2008).

Figure 12. Condition categories define as Q´s for Costa Rica´s National Paved Road Network.

Reports released by LanammeUCR not only bring data on pavement condition but also on the investment done on them. Since 2013, the information derived from each road work was included for the principal activities that contractors carry out (Hot Asphalt Mix, Potholing, Crack Sealing, among others). It should be noted that there is no automated or modern payment control and recording system in the Road Authority in Costa Rica, so maps like the one in the Figure 14 are a valuable output from the oversight model of LanammeUCR.

Considering the investment information gathered in that data base for each assessment report, it was possible to create an investment efficiency index by LanammeUCR. In a simple way, IRI and FWD information of each section of the national paved routes are compare from one assessment to another to see the condition variation. That change in the pavement condition (improvement, deterioration or no variation) is also compared to the investment done in each section to determine the impact of the resources spent in the road works.

The Figure 15 presents the information about the performance of paved road sections after the investment performed in each road section.

The oversight system designed for the case of Costa Rica´s National Road Network executed by LanammeUCR is also an important source of information to develop research on pavement management. More than ten research projects have been presented on this field, considering topics like deterioration models, strategic and tactical investment plans, strategic pavement condition indexes, institutional practices gap analysis and municipal level pavement management.

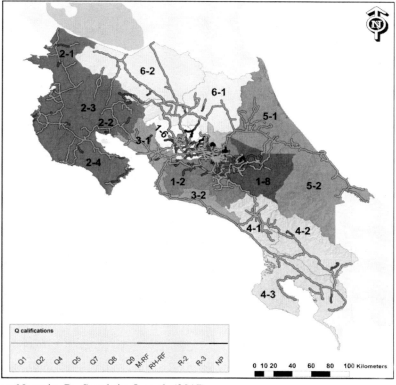

Source: Naranjo, R., Sanabria, J. et al. (2017).

Figure 13. Map of attention strategies for each road network section according to Q's.

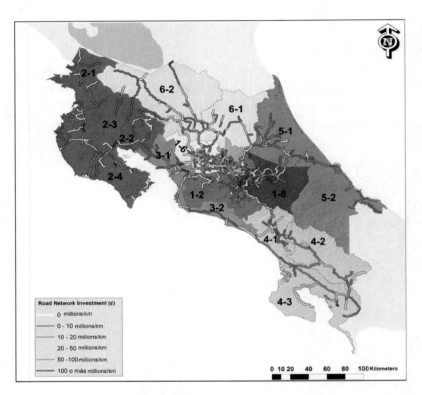

Source: Naranjo, R., Sanabria, J. et al. (2017).

Figure 14. Investment performed on paved road network in Costa Rican colonies 2016-2017.

CONCLUSION

Many countries and states have shown that Pavement Management and Transportation Asset Management principles work. There is strong evidence about the benefits in terms of performing a more efficient investment when these principles are followed. Even in the cases where not all the components of a Pavement Management System are implemented, the benefits are perceived and obtained.

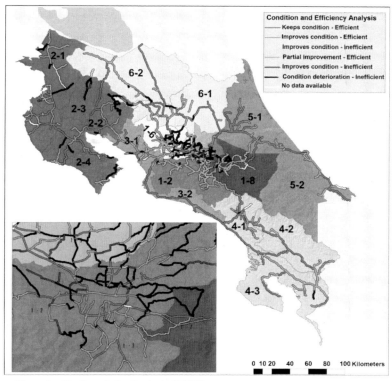

Source: Naranjo, R., Sanabria, J. et al. (2017).

Figure 15. Efficiency index in Costa Rican National Road Network.

Costa Rica is an example of that because, in despite of the lack of an institutional Pavement Management System implemented, what has been accomplished through a road sector oversight model that enables a public university to participate in the sector, shows the potential of a PMS. It is also a special situation because of all the available information of seven paved national road network assessments.

One cannot forget that implementing a PMS is a matter of change. As a change, there will always be opposition. Resistance to change is inherent to asset management implementation processes and of course, there will be resistance when it comes to pavements, the largest asset in a transportation system.

As an example, in the Costa Rican experience, the incorporation of IRI as a parameter of acceptance in road contracts faced discussion and resistance as is expectable. The key is to stay focus on the performance goals for the road network in the long term and use international experiences to demonstrate that other the countries faced the same doubts and they were overcome.

Technical strictness in road engineering is also a key to develop an oversight model like the one LanammeUCR carries out. This model is based on the asset management principle "keep it simple before fancy". That is recognizable in decisions like the one made for the first assessment in 2002, where the evaluation in a partial part of the paved road network was used to calibrate the equipment.

Finally, accountability is a central concept derived from PMS and TAMS in general. For the case of Costa Rica, it is the reason why the University of Costa Rica through LanammeUCR received a mandate from the legislators. Public universities are the most trusted institutions in Latin America, that is why they have been assigned the work to generate transparency, publicity and divulgation to road work investments. Good news is a PMS is key to generate trust in the public opinion, using principles like long term view, investment scenarios comparison, and data based decision making process.

REFERENCES

AASHTO (2011). *Transportation Asset Management Guide*. U.S. Department of Transportation.

Federal Highway Administration (1999). *Asset Management Primer*.

LanammeUCR. (2008). LM-GI-EV-01-2009 *Informe de Evaluación de la Red Vial Nacional Pavimentada de Costa Rica Año 2008*. [LM-GI-

EV-01-2009 Evaluation Report of the National Paved Road Network of Costa Rica 2008].

LanammeUCR (2006). *Informe de Evaluación de la Red Vial Nacional Año 2004.* [*Evaluation Report of the National Paved Road Network of Costa Rica 2006*].

LanammeUCR (2004). *Informe de Evaluación de la Red Vial Nacional Año 2004.* [*Evaluation Report of the National Paved Road Network of Costa Rica 2004*].

——. "LanammeUCR." 2014. http://www.lanamme.ucr.ac.cr/index.php/productos/cat%C3%A1logos/catalogo-equipos.html (accessed 11 6, 2017).

OECD (2001). *Ageing and Transport Mobility Needs and Safety Issues.*

OECD (2001). *Asset Management for the Road Sector.* Report.

Rodríguez-Morera, José D (2012). *Tesis Plan de inversión a nivel estratégico en pavimentos felxibles de la red vial nacional de Costa Rica.* [*Thesis: Strategic Investment Plan on asphalt pavements of the National Paved Road Network of Costa Rica*].

Sanabria, J., Barrantes, R. and Loría-Salazar, L. (2015). *INF-PITRA-001-2015 Informe de Evaluación de la Red Vial Nacional Pavimentada de Costa Rica 2014-2015.* [*INF-PITRA-001-2015 Evaluation Report of the National Paved Road Network of Costa Rica 2014-2015*].

Sanabria, J., Barrantes, R. and Loría-Salazar, L. (2011). *LM-PI-UE-05-11 Informe de Evaluación de la Red Vial Nacional Pavimentada de Costa Rica 2010-2011.* [*LM-PI-UE-05-11 Evaluation Report of the National Paved Road Network of Costa Rica 2010-2011*].

Sanabria, J., Naranjo, R., et al. (2017). *INF-PITRA-002-2017 Informe de Evaluación de la Red Vial Nacional Pavimentada de Costa Rica 2016-2017.* [*INF-PITRA-002-2017 Evaluation Report of the National Paved Road Network of Costa Rica 2016-2017*].

Sayers, M. and Karamihas, S. (1998). *The Little Book of Profiling.*

Sayers, M. and Karamihas, S. (1996). *Interpretation of Road Roughness Profile Data.* Final Report for FHWA.

Solminihac, H. (1998). *Gestión de Infraestructura Vial.* [*Road Infrastructure Management*].

Transportation Association of Canada (1997). *Pavement Management and Design Guide.*

World Bank (2007). *Data Collection Technologies for Road Management.*

In: Transportation Infrastructure ISBN: 978-1-53614-059-0
Editor: S. Antonio Obregón Biosca © 2018 Nova Science Publishers, Inc.

Chapter 4

AIRPORT PAVEMENT MANAGEMENT

Mauricio Centeno, PhD
Soluciones e Ingeniería en Vías Terrestres S.A. de C.V.
Guadalajara, Jalisco, México

ABSTRACT

This chapter focuses on the definition and brief description of the basic concepts related to Airport Pavement Management System (APMS). It shows the components that integrate an APMS, their relationships and how they interact to reach the objective of these systems, an optimum Pavement Maintenance Program. Finally, this document includes some ideas of future challenges in this topic.

INTRODUCTION

A Pavement Management is a system that uses a computer program to assist decision makers to develop the best strategies to maintain the pavements in good (acceptable) condition over a specified period for a user selected budget. APMS is a Pavement Management System (PMS) used in airport facilities.

A Pavement Management System provides an objective, schematic and well-defined system which use information of pavement condition over the time. Information on pavement condition, work costs, agency policies, etc. are stored in a database. The database is the core of the system, and it saves all the information with which the system works.

PMS' is relatively new. Many agencies since early 1980´s utilize these systems. Their development has gone hand in hand with the evolution of computer systems. The enormous increase in the processing capacity of the computers, make possible not only to generate PMS with more significant capabilities, but also it has been possible to make more detailed pavements evaluations. For example, nowadays is common to find PMS that include high-performance technology, such as HD video (see Figure 1).

Pavement evaluation information together with precise criteria provides the PMS user the possibility to develop a specific maintenance program for a pavement network.

In fact, the works for periodic condition pavement evaluation depends on the needs of the pavement manager (agency). It is possible to implement a PMS only with a visual inspection or with the use of high-performance equipment or also with a combination of both.

Figure 1. Examples of high performance technology equipment use for airport pavement evaluation.

PROJECT AND NETWORK LEVELS FOR PAVEMENT MANAGEMENT

There are two levels well-defined for Pavement Management. It is essential to be aware of the difference between these two levels, fundamentally to understand the results obtained from each of these levels.

Network Level Management

The first one is Network Level. It refers to global pavement management. Network Level is useful for defining the annual budget for a period of analysis for a complete pavement network. The expected result is a pavement maintenance program which includes the type of maintenance work required in each section of the pavement network during a given period. PMS works at the network level.

Project Level Management

The second is Project Level. It refers to a particular analysis to define the best option to solve a specific problem in a pavement section. In a project-level work, the engineer develops a solution for the current situation and a particular pavement section. In Project Level, the rest of the sections in a pavement network do not matter, neither does the past condition of the pavement; it only matters to find the optimal solution for the present moment. The expected result is a detailed project to solve a specific problem in the current condition of the pavement. The plan must include a description, material specifications, detailed topography, construction equipment needs and specifications, etc.

In fact, these two levels are complementary, that is, the results of the other level cannot substitute the effects of one level. A typical error is when an agency implements a PMS and only considers the network level management.

AIRPORT PAVEMENT MANAGEMENT SYSTEM COMPONENTS

In general, an Airport Pavement Management System integrates different components (see Figure 2), there are:

- Inventory
- Maintenance work history
- Pavement evaluation (periodically)
- Maintenance strategies, policies, and priorities
- Maintenance work costs
- Prediction models for pavement performance
- Analysis module

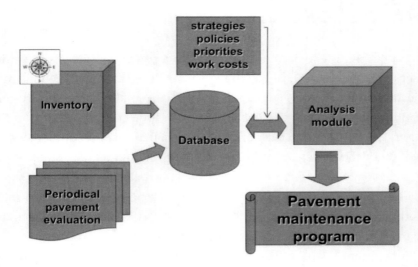

Figure 2. Graphic representation of PMS components.

All these components work together in a computer system to reach the best technical-economical pavement maintenance plan. Of course, if the APMS uses a computer system, all the information need to be organized in a database and the policies need to be clear.

All the APMS components are important to assist system users to get the best result, however it is possible to have a system without one of these components, although the results on the system may not be as precise as desire.

In the following subtitles there is a brief description of the components of a APMS.

Inventory

The first step in the implementation of an APMS is the definition of pavement network in the system, that is that commonly named as "Inventory." The definition of the pavement network implies first creating a hierarchy. The most common hierarchy is networks, branches, and sections. Paver®, a very used software for Airport Pavement Management uses this arrangement. As Shahin, M.Y. defines a section is "the smallest management unit when considering the application and selection of major maintenance and repair (M&R) treatments." Every section must have a name and a unique code for identification. In fact, M&R proposals stand on this definition. An incorrect interpretation of sections would cause a deficient pavement maintenance program.

Section

A section is integrated by pavement areas that share the same type of pavement (asphalt or concrete), use, structure, traffic, and repair history. Creating a good definition of the pavement sections makes possible to apply the M&R proposals in the right place, at the right time.

Branch

According to Shahin, M.Y. a branch is defined as "a readily identifiable part of a pavement network and has a distinct use" from the rest of the pavements. A branch or pavement use is easy to define in an airport network. Typically, the following branches are defined: runway, taxiway, apron, heliport, parking, and roads. A group of sections makes a branch.

Network

A network is a group of pavements that are managed by a single entity (agency/airport administrator). An individual bag is the funding of the maintenance work of a network, that is a separate entity manages funding sources. Usually, the airport pavement management creates one or two networks. When the two of them are designed, is typical to assign one for the operational area and the other one for the non-operational area.

Figure 3 shows an example of sectioning of an airport pavement network.

Reference System

Another vital aspect to consider in an inventory definition is a reference system to locate all sections of the pavements in its network. These requirements are represented in Figure 2 with a wind rose. Nowadays, with the use of Global Positioning System equipment (GPS) or a Geographic Information System (GIS) is easy to create a coordinate system. In Figure 3 the airport pavement network is represented in a GIS, that helps to find easily every section of a plan. This representation of the pavement network is handy for in-place works, like a periodical evaluation of pavements.

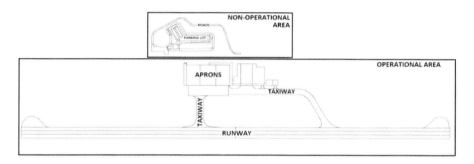

Figure 3. Graphic representation of airport pavement network.

Maintenance Work History

This component of APMS is essential but not necessary for the system. A Maintenance Work History integrates all the information listed below:

- Work type (preventive or corrective)
- Work detailed description (volumes, materials, thickness, and number of layers)
- Date
- Contractor
- Cost
- Pavement Section

With the Maintenance work history information, is possible to analyze if the works have the expected impact on pavement condition. Furthermore, this information is essential for the calibration of the system. In other words, with the work history analysis, you can get the "lessons learned," and this makes it possible to adapt the system to local conditions.

Also, having all the maintenance work history information helps to evaluate other aspects such as the performance of the contractors, the functioning between the use of different materials, machinery or techniques for the same work and having easy access to historical information.

In general, the maintenance work history analysis is helpful to get the most benefit from the pavement management regarding system calibration and improvement of maintenance works.

Pavement Evaluation

This process is fundamental for an APMS. With this process an APMS determines the information of pavement condition over time (past, present, and future). There is not a unique way to evaluate a pavement, it is possible to select what kind of evaluation should be executed and the frequency of assessment. A common practice in airports is to perform an evaluation once a year because it is an adequate period to notice clear differences between two consecutive measurements and the economic resources can easily be programmed to fund the expenses of each measure.

Standard practice in airport pavement evaluation is to perform a visual inspection. Usually, the visual inspection for airports follow the process described in ASTM Standards D5340 and D6433, Standard Test Method for Airport Pavement Condition Index Surveys and Standard Practice for Roads and Parking Lots Pavement Condition Index Surveys, respectively. The result of the visual inspection following the ASTM standards mentioned above is the Pavement Condition Index (PCI) for each section of the pavement network.

Pavement Condition Index

PCI is a numeric index ranging from 0 (for a complete failed pavement) to 100 (for a new pavement). The PCI is calculated based on the type, the degree of severity and area affected by distresses present on pavement section. Then, pavements with many or more severe distress will have a lower PCI than pavements with less deterioration. The Paver® software uses PCI as its parameter for determination of pavement condition.

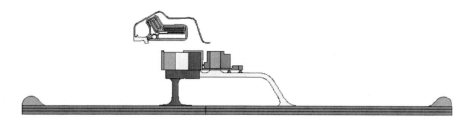

Figure 4. Graphic representation of PCI for an airport.

With the use of the PCI and GIS, it is possible to graphically represent the condition of the pavement at a certain point in time. As shown in Figure 4 each color on the map represents the state of the pavement section. It is possible to have one map per year to analyze the evolution of PCI in an airport pavement network.

International Civil Aviation Organization (ICAO) and United States Federal Aviation Administration (FAA) recommend for safety and fatigue on airplane structure the evaluation of evenness on a longitudinal profile (also called Pavement Roughness) and friction in the runways.

Runway Pavement Roughness

In document AC No. 150/5380-9, Guidelines and Procedures for Measuring Airfield Pavement Roughness, FAA gives it recommendations to evaluate evenness on runways. As it mentions on the previous paragraph, irregularities on pavement surface cause a lack of comfort for passengers, can diminish safe on operations, increase fatigue on airplane structure and the possibility to the constitution of water pools that can induce aquaplaning. From all points of view, it is desirable to have fewer irregularities as possible irregularities in the pavements. For that reason, there are some price adjustments agencies have begun to apply based on roughness the contractors leave on their work. In Table 1 there is an example of the criteria proposed in FAA document AC 150/5370-10G, Standards for Specifying Construction of Airports.

Table 1. Example of price adjust factor definition based on roughness (taken from AC 150/5370-10G)

Average Profile Index (Inches Per Mile) Pavement Strength Rating			Contract Unit Price Adjustment (PFm)
Over 30,000 lb	30,000 lb or Less	Short Sections	
0 - 7	0 - 10	0 - 15	0.00
7.1 - 9	10.1 - 11	15.1 - 16	0.02
9.1 - 11	11.1 - 12	16.1 - 17	0.04
11.1 - 13	12.1 - 13	17.1 - 18	0.06
13.1 - 14	13.1 - 14	18.1 - 20	0.08
14.1 - 15	14.1 - 15	20.1 - 22	0.10
15.1 and up	15.1 and up	22.1 and up	Corrective work required

Unfortunately, the irregularities are increasing with the use of the infrastructure, the only way to reduce them is through a work of corrective maintenance. In asphalt pavements (AC) the restorative work is cold mill and replace of surface asphalt layer; for hydraulic concrete pavements (PCC) the corrective action usually applied is surface grinding.

There are many options for equipment that measures pavement roughness, and the most common are profilers, inertial profilers and response type equipment. Also, there are many profile indexes to evaluate pavement roughness, like Profile Index (PI), International Roughness Index (IRI) or Boeing Bump Index (BBI). Depends on each agency which equipment and index use.

Runway Pavement Surface-Tire Friction

The friction between tire and pavement surface is the other parameter evaluated in runways by safety issues. Friction is necessary because two aspects: 1) to avoid the misdirection of the aircraft towards lateral part of the track by sliding during the landing or take off operations and 2) to guarantee a sufficient braking distance for the plane during landing operations.

Table 2. Friction levels on ICAO Annex 14 Volume I

Test equipment	Test tire Type	Pressure (kPa)	Test speed (km/h)	Test water depth (mm)	Design objective for new surface	Maintenance planning level	Minimum friction level
(1)	(2)		(3)	(4)	(5)	(6)	(7)
Mu-meter Trailer	A	70	65	1.0	0.72	0.52	0.42
	A	70	95	1.0	0.66	0.38	0.26
Skiddometer Trailer	B	210	65	1.0	0.82	0.60	0.50
	B	210	95	1.0	0.74	0.47	0.34
Surface Friction Tester Vehicle	B	210	65	1.0	0.82	0.60	0.50
	B	210	95	1.0	0.74	0.47	0.34
Runway Friction Tester Vehicle	B	210	65	1.0	0.82	0.60	0.50
	B	210	95	1.0	0.74	0.54	0.41
TATRA Friction Tester Vehicle	B	210	65	1.0	0.76	0.57	0.48
	B	210	95	1.0	0.67	0.52	0.42
GripTester Trailer	C	140	65	1.0	0.74	0.53	0.43
	C	140	95	1.0	0.64	0.36	0.24

In Annex 14 to the Convention on International Civil Aviation Volume I, Aerodrome Design and Operations, Table A-1 shows friction levels for new and existing runway surfaces. This table is reproduced in Table 2.

In Table 2 is shown the friction levels at different measure speeds (65 and 95 km/h), at different phases of pavement life (new surface, maintenance level and minimum/end of pavement life) and different water deep applied. Besides, Table 2 shows the test equipment most used to measure friction on runways. The configuration of each measure friction equipment is different from one another; however, the levels for each of them are available.

Maintenance Strategies, Policies and Priorities

The airport administrator defines the strategies and policies for maintenance. The policies define the rules for maintenance work to be accomplished for any situation/condition of pavements. One of the most common policy refers to create a relation between maintenance works and current distress on the pavement. Of course, there is not a unique work for specific distress; each airport policy will decide what action to apply for any distress. Even more, within an airport, you can have different policies per network (in the case of having two or more). Naturally, each airport has their policies depend on local conditions (materials, contractors, pavement type or their state of practice).

Figure 5 shows an example of how Paver® software selects a maintenance work within policies. In this example, there is selected an AC Deep Patching (maintenance work) for High Severity Alligator Cracking (current distress), but it could choose another maintenance work according to technicians' criteria.

With well-defined policies, the APMS can determine the volumes and costs to be executed year after year, depending on the condition of the pavement through the time.

With clear-cut policies, the APMS can determine the volumes and costs to be executed year after year, depending on the condition of the pavement through the time.

The strategies represent the way in which the airport administrator pretends to carry out pavement management to fulfill the objective of having the pavements in the best possible condition with the available economic resources. One strategy, for example, could be to apply only to extensive rehabilitation (corrective maintenance), instead of also applying preventive maintenance or the use only AC instead of PCC for runways. There are many strategies for airport administrators, the most critical issue is to implement, perform and evaluate the plan to determine if it is adequate for the airport. A good practice is to assess the results (cost vs. benefits) every certain period (for example 5 years), to continue with selected strategy or if an adjustment is necessary.

In general, the funds for pavement maintenance are insufficient for all pavement maintenance needs in an airport. The administrator has to decide which pavement sections are more critical for operational requirements. These sections have to be taken care of before the minor ones. In general, for airport pavements is so easy and obvious to establish the priorities. The most crucial pavement sections are those that are in the operational area. Inside operational areas, the first to be cared for should be runways, then

Airport Pavement Management 115

taxiways and aprons. In non-operational areas, roadways, and parking lots for passengers should be the first to have cared before facilities for airport employees. APMS uses rankings to establish priorities, based on the type of traffic, the category of a section, type of aircraft, etc.

The experiment people should establish the strategies and priorities in a multidisciplinary group of experts. These represent the component of the system that necessarily requires human intervention since it is a decision-making process.

Maintenance Work Costs

Each airport administrator decides what maintenance work will be carried out on their pavements based on the technology and materials available in their area. This decision seems very logical; however, it is not always so easy. There are airports where the quality of the aggregates for the production of AC or PCC does not meet the standards. For example, in some regions of Florida is very difficult to find polish-resistant aggregates. The lack of the polish-resistant aggregate could be a big problem due to it being difficult to meet the friction value requested in the standard shown in Table 2. Or other issues such as not having an AC production plant near to airport or in coastal areas, the high humidity makes almost impossible to apply slurry seal at night works, etc.

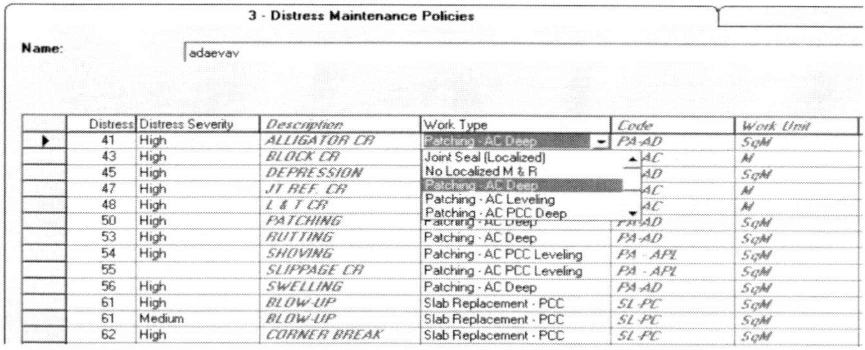

Figure 5. Example from work policy define window in Paver® software.

Local conditions make it necessary to reduce the available alternatives of maintenance works already exist. But the reduction of the work options is not the only problem; another perhaps more significant issue is the impact of those conditions on the cost of the works. For example, the hauling of aggregates to regions like Florida could increase the cost of works significantly.

Then, when using an APMS, it is necessary to take into account the local conditions for the definition of the works to be executed and the costs associated with them. The constant updating of costs (once a year minimum) is necessary for the PMS to deliver adequate results. Technicians are also recommended to stay updated regarding new techniques for airport maintenance works.

Prediction Models

These models have been developed for years by experts. The specific equations to be used will be taken, depending on the APMS selected and the pavement evaluation parameters selected. In fact, almost all pavement management systems include that equations in its software. For example, Paver® software uses polynomial equations to approximate the behavior of the PCI through the time. Figure 6 shows an example of PCI prediction model for Paver® software.

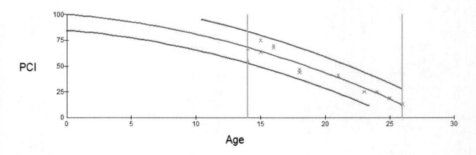

Figure 6. Example of PCI prediction model in Paver® software.

The most critical point of the use of the prediction equations is to assure they represent the behavior of the pavement over time in all the values range. It means the prediction values of the system must be very close to the future values obtained with the evaluations. Of course, it is necessary that the equations are calibrated through the time, considering the values obtained during the pavement evaluations.

It is recommended to define one equation at least for each branch established in the pavement network. Also, is recommended for evaluation parameters like BBI, PI or Friction to measure all the variables used in the equation, that is to avoid as much as possible the use of default values in equations. If there is not possible to measure all parameters, it is better to change the equation or system.

Analysis Module

This element of APMS integrates all the information of the rest of the components. Use the information in the database to obtain future values of evaluation parameters with the prediction models. With present and future values of the evaluation parameters (like PCI), inventory, strategies, policies, priorities, maintenance works and their costs; the analysis module can calculate the volumes (and expenses) of work required for the period of analysis established.

This module has many inputs and uses all of them to create a Pavement Maintenance Program according to the needs indicated by the system user. Typically, these needs could be a constraint budget or budget needs to reach some pavements condition. Each set of requirements is called a scenario.

Usually, the APMS is used to determine the pavement maintenance program for various scenarios. Subsequently, the decision makers select which program will execute. The authorities will have to approve the program selected. Pavement Maintenance programs are usually made for at least five years.

Conclusion

The development of airport pavement management systems has had a significant advance over the past 20 years. Among other things, it took place because, in 2001, due to the implementation of a Safety Management System as a requirement for Aerodromes Certification.

However, some challenges have arisen with the use of systems. Next, there are some ideas that this author wants to leave embodied in this document, which is the result of personal experience and documentary research.

- It is vital that the authority determines a single method/equipment for measuring the pavement evaluation parameters, in concrete for the roughness and friction on runways. The use of several alternative methods that measure the same specification is somehow ambiguous when it is possible to measure with different equipment since it is probable to be within the standard value with one equipment and outside with another. This situation may be due to a commercial issue more than to a technical problem.
- The system update is a continuous activity that must be carried out. This update refers to all the components of the APMS from the inventory, priorities, costs, strategies and prediction models. An Airport Pavement Management System loses effectiveness the more obsessed is the information it contains.
- An APMS is not only a requirement for airport certification. It is a planning tool that makes the investment in the maintenance of pavements more effective. Some administrators tend to see it as a useless mandatory requirement, but it is not. It is essential that all administrators (agencies) give the system an opportunity to demonstrate it works and it generates savings.
- The APMS together with good executive projects (Project Level) along with a proper execution are the best way to make the investment profitable, with this we make sure appropriate use of the funds invested in pavement maintenance. It is necessary to

carry out the three activities as good as possible, so if one of them fails, probably the other will not work efficiently.
- It is not possible to request and obtain the economic resources in a better way, than technicians showing to the financial areas the economic consequences of different scenarios (budgets) for an airport pavement maintenance program. With an APMS tool is easy to explain and easy to understand for anyone.
- Another challenge is to develop APMS that use as much testing and processing equipment technology as possible, seeking to reduce the costs of their operation and deliver the pavement management programs in shorter periods. In such a way that they are more attractive for airport managers.

REFERENCES

American Society for Testing and Materials (1992). *STP 1121, Pavement Management Implementation*: 228-272.

American Society for Testing and Materials (2011). *ASTM D 6433, Standard Practice for Roads and Parking Lots Pavement Condition Index Surveys*: 1-8.

American Society for Testing and Materials (2012). *ASTM D5340, Standard Test Method for Airport Pavement Condition Index Surveys*: 1-9.

Federal Aviation Administration (1997). *AC No: 150/5320-12C, Measurement, Construction, and Maintenance of skid-resistant Airport Pavement Surfaces*: 1-28.

Federal Aviation Administration (2014). *AC No: 150/5380-6C, Guidelines and Procedures for Maintenance of Airport Pavements*: 1-45.

Federal Aviation Administration (2014). *AC No: 150/5380-7B, Airport Pavement Management Program (PMP)*: 1-18.

Federal Aviation Administration (2016). *AC No: 150/5320-6F, Airport Pavement Design and Evaluation*: 1-110.

Federal Highway Administration (2006). *ASTM D 6433, Standard Practice for Roads and Parking Lots Pavement Condition Index Surveys*: 1-8.

Niju, A. (2006). *GIS based Pavement Maintenance & Management System (GPMMS)*: 1-24.

Shahin, M. Y. (2002). *Pavement Management for Airports, Roads and Parking Lots, Second Edition*: 1-252.

Transportation Research Board (2008). *Circular No. E-C127, Implementation of an Airport Pavement Management System*: 1-17.

Transportation Research Board (2011). *ACRP Synthesis 22, Common Airport Pavement Maintenance Practices*: 1-40.

In: Transportation Infrastructure
Editor: S. Antonio Obregón Biosca
ISBN: 978-1-53614-059-0
© 2018 Nova Science Publishers, Inc.

Chapter 5

EFFECTIVE ASSESSMENT AND MANAGEMENT OF RAILWAY INFRASTRUCTURE FOR COMPETITIVENESS AND SUSTAINABILITY

José A. Romero Navarrete[1,]*, Frank Otremba[2] and Saúl Antonio Obregón Biosca[1]

[1]Queretaro Autonomous University, México
[2]Federal Institute of Materials Research and Testing (BAM), Berlin, Germany

ABSTRACT

The inherent lower cost and environmental impact make railways a highly competitive transport mode for both passengers and cargo. However, such competitiveness can be affected by poor assessment and maintenance practices. The different principles considered for keeping in good shape the infrastructure, include preventive, corrective and predictive maintenance approaches. Such different preservation techniques have been considered as vital tools to fulfill the final mission of a railway infrastructure, which is to provide economy, safety and reliability to its users. In this respect, advanced infrastructure monitoring systems have been created for some highly demanding railway transport systems, in the form of active systems that aim at the prevention of excessive damages caused to the infrastructure by the rolling stock. Such systems include the rail stress, in order to detect any defect in the railway supporting system. The competitivity challenges of this transport are illustrated in this chapter by two separated situations, involving accidents in Canada and Germany. Future actions are identified for stablishing rational railway pricing, as a function of the railway damage potentials of the different payloads on the cars.

Keywords: transportation infrastructure, railway transportation, assessment, management, modelling

INTRODUCTION

Running an enterprise dedicated to operating the railway transportation infrastructure of a country or region, encompasses many activities that keep different levels of closeness with the final mission of such type of enterprise: to provide its users, a reliable and safe transportation infrastructure. In this sense, there are support activities and substantive activities. While support activities include finance, legal issues and accounting operations, substantive activities of such companies include the management of the infrastructure maintenance operations, and the assessment of the condition of such assets. In this context, the aim of any railway infrastructure maintenance activity, is to keep such assets in a condition which represents the minimum costs to railway users, and involves the minimum safety risk. For a railway infrastructure, economic operational conditions include the following:

- *Good ride quality,* that is, the roughness of the rail and the whole railway structure, should not cause excessive vibration to railcar´s payload and railcar components. Cargo can be damaged due to vibrations
- *Minimum delays* due to railway maintenance operations

- *Minimum disruptions* due to railway maintenance operations.

On the other hand, safe operational conditions of a railway infrastructure, imply that the trains, under normal operation, can circulate without any exposure to suffer accidents. Some safety characteristics are listed below:

- No derailments conditions due to broken or over-wearied rails
- Acceptable wheel-rail friction properties to perform accelerating-decelerating maneuvers.

Assessment activities in the context of railway infrastructure, involve its characterization, to learn about how apt it is to provide the intended service at a certain cost. Accordingly, assessment operations include the following:

- Measuring the economic viability of the infrastructure, that is, to establish what is the ratio of the maintenance investments in relation to the revenue
- Measuring the level of service provided to the railway users, characterized by different indicators.

For stablishing the preservation investment policies in railways, the complexities of the economic relationships between the railway users and the infrastructure operators, should be anticipated. For example, poor riding conditions can induce vehicle components deterioration, which can generate a poor service to the final users in the case of passenger trains.

The railway infrastructures are in general, evaluated with different metrics to describe its condition, in a context where different performance measures must be managed in order to avoid any increased damage of the infrastructures as a result of applying delayed corrective measures (Gaudry et al., 2016). The management of the railway infrastructure operations, involve a complex network of mutually influencing characteristics and parameters. In this respect, on the left side of the diagram included in

Figure 1, the concepts that are under control of the management are described. The core of this diagram is the physical infrastructure, whose performance can be affected by the administration, on the left side, but also by an extraordinary demand and/or by the environment. On the other hand, and on the un-controlling situation for the management, there are environmental effects, that can include the presence of nature phenomena, such as hurricanes and earthquakes.

In the diagram of Figure 1, the different concepts behind the maintenance operations, are described, in terms of the type of infrastructure maintenance criteria. In general, there are two criteria to carry out the maintenance activities: preventive and corrective maintenance. While the corrective maintenance involves a failed situation that needs to the repaired, the preventive criteria denotes the avoiding of any failure in the infrastructure components, in such a way that the consequences of failures, are avoided. However, a variety of situations can be identified between these extreme maintenance criteria. For example, as it is illustrated in this diagram, the preventive maintenance can be systematic, conditional or provisional, as a function of the timing and inspection procedures, where the timing can be defined in terms of hours of service or mileage travelled. The systematic maintenance is based upon fixed timing/use criteria concerning the replacement of the parts and materials; the conditional preventive maintenance deals with the replacement of parts and materials before they fail, but that exhibit signs of deterioration. Consequently, such conditional preventive maintenance principle depends on the inspection of the parts and materials, thorough a component condition monitoring scheme, representing an economically attractive concept. The benefits of applying the preventive maintenance criteria, include that no un-expected breakdowns occur, as the maintenance operations can be scheduled. In this context, provisional maintenance refers to no-permanent reparations, made to avoid interruptions in the transport service.

The maintenance principles that are considered for preserving the integrity of the infrastructure while having acceptable levels of functionality, are related to general concepts, common to other industries, including the different types of maintenance, mentioned above. Selecting the appropriate maintenance methodology, however, is a cumbersome activity. According to some organizations, a certain ratio between both types of approaches, as far as the cost and time involved in such operations, can be established (Stenström et al., 2016).

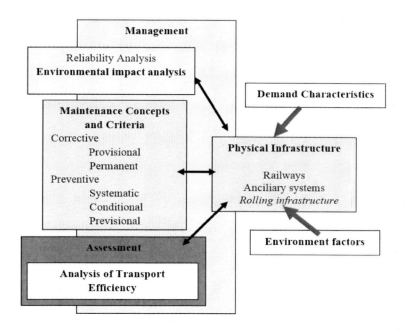

Figure 1. Management activities for railway infrastructure.

In this context, assessment is considered as the activity that aims at evaluating the level of efficacy and efficiency of the service provided by the infrastructure to its users. As a result of this assessment, different measures can be taken by the management, including the undertaking of maintenance operations, repairing and even overhauling (MRO) (Rodrigues and Lavorato, 2016). A multitude of criteria is involved when making the decision for applying any of these different criteria to reestablish the reliability, economy and safety of the transport system. The outputs from this assessing process can thus consist of working orders and requirements that restrict, for example, the amount of resources allocated for a certain infrastructure.

The characteristics of the infrastructure assessing process, derives from estimations of the expected market for the transportation system. That is, the assessment must integrate forecasting of the level of usage of the infrastructure. The estimation of the demand for a certain infrastructure thus sets the investment limits for maintaining such infrastructure at the needed economic levels of operation and safety. An effective assessing of the infrastructure involves background, present and forecasting information, integrating data about the conditions under which the infrastructure has been used; the present condition; and the forecasting of the infrastructure´s service decay. This process thus involves data handling that allows the forecasting of the evolution of the system. Figure 2 depicts a scheme of these relationships between operation/maintenance data and assessment of the infrastructure.

As for any other field of transportation, technology plays a crucial element for having an efficient and effective transportation system infrastructure.

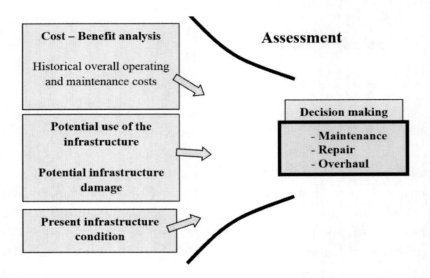

Figure 2. Schematic diagram of the inputs/outputs of the assessment activities.

For example, the traditional bare eye inspections have been replaced by automatic Laser systems of inspection that acquire data about the shape of the rails, and reports directly to the central stations through internet.

Asset management in the case of railway networks covers a multitude of activities that influences the overall performance of any´s nation economy. Poor asset management can thus signify for a country higher costs for the products and a lower competitivity in a global environment, in which manufacturing components travel across borders several times before reaching their final configuration.

It is important to note that the management systems that provide the best solutions for assets management, are the ones that are considered since the very initial stages of the infrastructure design.

Furthermore, a reliable measure of the condition of the infrastructure is critical for the economic and timing realization of the maintenance and preservation activities. In this respect, in Europe, the initiative "Infrastructure Needs Assessment Tool" was implemented in the Central European Region, aiming at extending the useful life of the infrastructure.

It is the purpose of this chapter, to review and to illustrate the range of technological possibilities for carrying out the maintenance operations of the railway infrastructure, in order to formulate specific technological needs for improving the quality of the analyzes and assessments. Also, to discuss specific examples of situations that pose competitivity and sustainability risks to railway transportation. The first part of this chapter, deals with the existing resources for infrastructure assessment, focusing on the available systems to perform a predictive maintenance. This section includes some simulation results of the rational behind some of these preventive maintenance systems. The second part of this chapter deals with the current research focus in this industry, identifying key research projects whose purpose is to improve the operation and working life of the infrastructure. The third part deals with the challenges facing the infrastructure management, which is illustrated through three examples. The fourth part of this chapter describes a study case of an which is highly relevant in the context of assessment and managing, that is, the Eurotunnel.

Finally, a series of conclusions and recommendations for future work, are described.

INFRASTRUCTURE ASSESSMENT IN THE RAILWAY INDUSTRY

The high operating costs associated to rail maintenance operations have motivated the development of innovative techniques and methods, aimed at monitoring the overall condition of both the rail and the rolling equipment. The resulting devices have involved the design of sensors, from weigh in motion devices to rail friction meters. All of these techniques for monitoring and surveying railways can be allocated within the concepts of a combination of preventive and predictive maintenance operations.

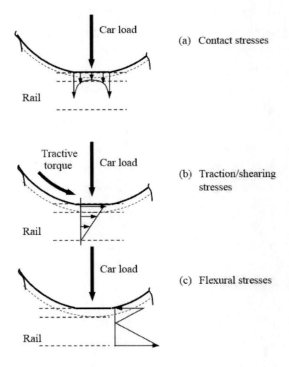

Figure 3. Diagrammatic representation of the stresses in the rail.

In general, rail material failure mechanism can derive from two effects: rolling contact fatigue and wear. While rolling contact fatigue depends on the magnitude of the combined stresses at the wheel-rail interface; wearing of the rail surfaces depends on the conditions of the wheel-rail interface, including the associated friction coefficient, the magnitude of the wheel force and the condition of the rail surface. Figure 3 illustrates a schematic representation of the level of rail stresses at the wheel-rail interface, where the stress level combines contact, traction and flexural stresses, involving a complex superposition of shearing and normal stresses.

The different predictive and conditional-preventive maintenance operations aim at: i) characterizing the conditions that would lead to an exacerbated rolling contact fatigue, which is mainly affected by the level of wheel forces; and ii) characterizing and preventing the amount of wear of the railway surfaces. The various technologies developed for characterizing these conditions, are described in the following paragraphs.

Weigh in Motion

Weigh in motion of railways cars target at the identification of operational conditions that can deteriorate railways at a higher than expected rate. In this respect, the common hardware for this monitoring consists of measuring the level of stress in corresponding points of the track, on the left and right sides.

Uneven wheel loadings can be along the longitudinal and/or cross car axis, as a result of miss -loading or -securing operations during the loading process of the railway car. Any imperfection in the railway car loading conditions can exaggerate the damage of the rail at specific wheel/rail contact points, as the potential effect of the wheel forces does not follow a linear relationship with respect to its absolute value. That is, according to the principles of contact mechanics, a certain increase, I, from the nominal load level, M, creates a reduction in the fatigue life of the infrastructure, on the order of the inverse of the $((I+M)/M)$ term, raised to a certain power (*fourth power law*). According to such validated relationship, an increase of only 10 percent in the wheel force, generates a reduction in the expected life of the pavement of 31.6%: *(1-(1/ (1.1)^4)) *100*. As an illustration of these phenomena, part (a) of Figure 4 illustrates some simulation results of the spatial distribution of the normalized-wheel forces along the external track of a curved track segment, as a function of the railway car`s speed. Part (b) of this figure illustrates the potential damaging effect of such force spatial distribution, according to the fourth power concept, for a traffic composed of railway cars travelling at different speeds and having different distances between bogies. According to these simulation results, there is a spatial determinism of the potential damaging effect of the wheel forces on the railway damage where, as an average, certain portions of the track can suffer up to 5.2 times more damage than other portions of the track.

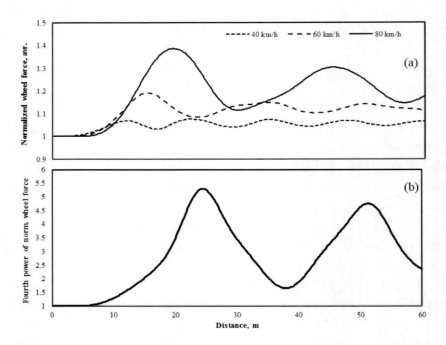

Figure 4. Simulation of the dynamic wheel forces and their rail damage potentials (four-power law).

These results thus provide an idea of the rail damage potentials of any un-even cargo distribution on the railway car, as the damage of the rail follows a nonlinear relationship with respect to the absolute value of the wheel forces. It should be noted that this uneven wheel forces can also be the result of moving cargos, such as swinging or sloshing cargoes.

In relation with the simulation results presented in Figure 4, it should be noted that such forces are also transmitted to the ground through the railway support components, in such a way that certain portions of these structural components, would be subjected to higher demanding conditions.

On the other hand, these results also suggest the negative effects on the railways, of any railway car overloading practices. Namely, weigh in motion systems can also be used to detect overloaded cars that can deteriorate rail material at a faster rate.

Railway Stress as a Function of Railway Support Stiffness

As it was mentioned above, the magnitude of the wheel forces defines, in a great extent, the rolling contact endurance of the rail material. However, it was also described that the stresses in the rail, are the superposition of at least three stress components involving shearing and normal stresses: wheel load, axle traction moment, and flexural stresses derived from the continuous support of the railway. The connection between railway's predictive maintenance and the level of railway flexural stress, results from the dependency that the level of such stress has on the stiffness rate of the railway supports. Namely, a weaken railway support means higher level of stress in the rail. This condition is illustrated in Figure 5, where the central sleeper is soft when compared with the adjacent sleepers. Such softening condition for the sleeper, can in turn be the result of the sleeper material fatigue; or the failure of the sleeper support itself. In this sense, Figure 6 illustrates simulation results of an in-house computer program to simulate rail's flexural stresses as a function of the softening level of the central sleeper in Figure 5 (Romero et al., 2016). According to these results, flexural stresses can double in the extreme case that the intermediate support completely losses its stiffness. Such higher levels of stress signify that the fatigue life of the rail will be reduced, as a function of the level of weakness of any intermediate support and of several operational conditions of the railways.

The rail stress measuring instrumentation thus provides information about the conditions of the railway support, and about the needs to restore / substitute any weakened supports.

The rail stress measuring instrumentation thus provides information about the conditions of the railway support, and about the needs to restore / substitute any weakened supports.

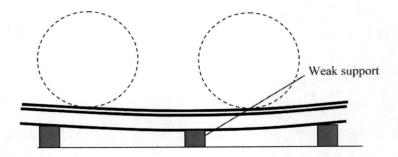

Figure 5. Representation of a failed/softened infrastructure support.

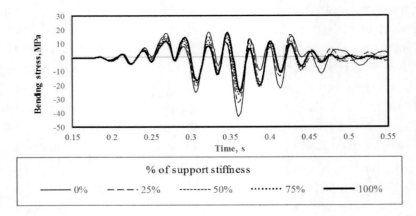

Figure 6. Stresses at the right node of the central infrastructure finite beam element, as a function of the normalized stiffness of the central support (Figure 5).

Railway Stress Variations as a Function of Railway Car Wheel Condition

In addition to characterizing the railway support capacity as a function of the railway stress level, the time variation of the rail stresses can be also indicative of the condition of the railway car wheels and axles. The axle and wheels defects that can be detected through rail stress measurement, include axle unbalance or eccentricity, and wheels flats. Axle/wheel unbalanced conditions, and wheel flats, represent a higher frequency vibration condition which can be sensed in the rail´s stress variation.

Friction Surveying

Accelerating and decelerating the train involves the application of corresponding traction or braking torques on the railway car axles, whose effectiveness depends on the available friction at the wheel-rail interface. Therefore, keeping high levels of friction is important for having both a reliable and a safe railway transportation, whose operation involves different levels of traction and braking torques. Measuring of the friction coefficient in a railway system, is important and it is associated to the capacity of the railway to provide the necessary friction when the vehicle performs speed changing maneuvers. For this purpose, there are several devices in the market, to perform such measurements.

Railway Wear and Failure Characterization

Monitoring of the wearing of the rail´s head and body is crucial for preventing major railway damage and potential broken rails that can lead to derailments. Also, to prevent derailments due to excessive rail distortion.
Automatic scanning systems have been developed to monitor the geometrical characteristics of the railway. Such monitoring systems can

operate with Laser, and are mounted on railway cars, specially designed to carry and operate such measuring devices.

CURRENT RESEARCH FOCUS

Research in the field of assessment, management and sustainability of the railway infrastructure, has had a range approaches to formulate solutions for increasing the rail transport competitivity. Such approaches cover from green logistics to applied research dealing with basic operation and maintenance issues.

Switches/Turnouts

An example of the basic technical problems that are being studied and analyzed in the field of railway infrastructure, is the issue of the effectiveness and effect of the switches/turnouts. These parts represent critical components for a railway system, and their maintenance is costly. The components of these devices, are subjected to impact forces as the train is changing its direction, and also involve the impact of the wheel when traversing the crossing nose of the turnout. Figure 7 illustrates such devices in a trail facility in California, United States.

Three situations are identified concerning such change of rail car direction (Doulgerakis, 2013 and Xin et al., 2016): i) the bladed shape of the re-routed rail, represents a weak component, susceptible to suffer a fatigue failure as a consequence of its reduced strength; ii) the maintenance of the switching mechanism, which is used for shifting the movable rail in this kind of switching mechanisms.

In this respect, it has been reported that the failure of the approaching blade-shaped rail segment, was the main cause for an important percentage of the disruption time in railway stations. The fatigue failure of this critical rail segment has been numerically simulated or improving its design, however, no validation has been provided for those studies, and the respective models developed have not been comprehensive. For example, in Nielsen et al. (2016) the level of stress and rolling contact fatigue in this component are simulated, taking into account even the railcar dynamics, but do not take into account the weakness of the wedge-shaped blade.

The second aspect listed above refers to the wearing and maintenance of the switching mechanism. In Boschert et al. (2017) a basic research project is described for simulating the physical principles for the operation of the railway switch point machine. The effectiveness of this mechanism is fundamental for the safety and endurance of both the rail and the railway car rolling components. In this long term project, called "Railway Switch Point Machine simulator" (RaSPoMas), the focus is on the kinetics of the mechanism, in particular, the power required to displace the moving components. The electrical power demanded for the operation of the moving components is used as an indicator of the condition of the point machine. The causality for any increase in the power demand is directly related to the condition of the mechanism, as such power depends on the existing corrosion, wearing and/or misalignment.

Figure 7. Rail switch at Train Station, San Francisco, California, USA.

Explosion of Hazardous Materials

As an example of the research activities that are being carried in the case of the railway transportation of transportation of hazardous materials,

Figure 8 illustrates an explosion testing carried out at the Federal Institute for Materials research and Testing (BAM, Berlin), aiming at characterizing the time for explosion of an oil railway tank car when subjected to a pool of fire, and the potential effects on the infrastructure.

Figure 8. Explosion testing of an oil railway tank car.

THE COMPETITIVENESS AND SUSTAINABILITY CHALLENGES

The degree of inherent economic and environmental benefits associated to the railway transport depends, in great extent, on the operational efficiency of the implemented system for managing the assessment and maintenance operations of the infrastructure. This is particularly relevant in the case of the transportation of dedicated hazardous substances. Three examples are presented below about the way that the competitiveness of the railway transportation can be poor or low, leading to users to consider alternate modes of transportation.

The Lac-Mégantic Tragedy

In Quebec, Canada, there was an extraordinary situation on July 6, 2013, in Lac-Mégantic, when a train carrying crude oil derailed and exploded due to the failure of the braking system, amongst many other contributing factors. 47 people lost their life in this event. As a consequence of such a tragedy, authorities in The United States and

Canada, have opted for increasing the levels of passive safety of the railcars carrying hazardous substances, including the following (DOT-117):

- Thicker railcar containers to improve their puncture resistance
- Reinforced thermal protection to prevent fluid explosion for a longer time in case that the container is in a pool of fire
- Full head shells against impact with adjacent rail cars
- Enhanced protection of valves

While these modifications in the specification of these vehicles can improve its passive safety, these alterations in the characteristics of the railcars, can affect its comparative advantages, as these measures represents, for example, an increase in the weight of the vessels, on the order of 28%. Additionally to these changes in the design of the railcars, the authorities are promoting the use of pipelines instead of railways, for carrying crude oil.

To avoid any impairment for the transportation of crude oils, innovative technologies have been proposed, including the discretization and solidification of the bitumen in order to diminish the probability of leakage and sinking in water, besides avoiding any potential explosion.

Rastatt Disaster

Another weighty event affecting the competitivity of the railway transport due to a damaged infrastructure, is represented by the disruption/close of a critical link of the *Rhine-Alpine Rail Freight Corridor*, between Rastatt and Baden-Baden (along the Basel-Mannheim leg), posing expensive logistic changes. This disruption was the result of the collapse of around 150 meters of railway under which a tunnel was under construction. Figure 9 illustrates the rail corridor in which this disruption took place, few hundred kms north from Basel, which is near the border of Germany with both France and Switzerland. As it can be seen in this figure, diverging of the train traffic would involve international operations that further posed logistic problems. It can be observed in this diagram that this railway path connected many important terminals on both extremes of the route, with freight traffic consisting of raw and intermediate goods, many of them critical for many industries.

The number of weeks that this railway link was broken was 51, costing around 102 million of USD, just to the logistic companies. Such disruption was recognized as a "Rastatt disaster," as only 150 freight trains were diverted, from the 200 total trains in that route. Maritime transport along the Rhine was considered for some consignments. The resultant wandering of the transport was costly and time consuming. The operations of at least 29 transport and logistic associations, belonging to six countries in the European Union, were affected by this disruption.

For the passenger railways, this disruption represented the use of substituted buses running every five to ten minutes, to bypass the affected zone.

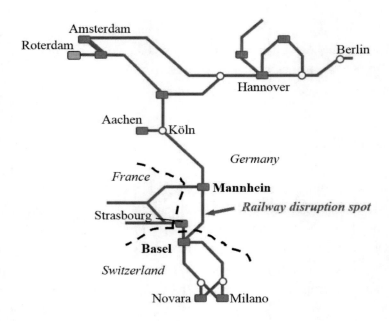

Figure 9. Rail disruption due to the collapse of the railway, between Rastatt and Baden-Baden, on August 12, 2017. The railway reopened till Oct. 2, 2017.

The Rastatt disaster unveiled a number of situations for infrastructure design and management, including the following (Berkeley, 2017):

- the absence of international crisis management tools
- the lack of viable, alternative routes
- lack of a common language for the railway operators in the region.

Electric Roads (ER)

A technology dating from the onset of the 20th century is being updated and conditioned for modern multi-lane highways. Such technology involves the transformation of internal combustion trucks into hybrid electric – fossil fuel trucks. Figure 10 illustrates a schematic representation of such a system, which is being implemented in a demonstrative project in Sweden.

The advantages of ER include the following (Connolly, 2017):

- It represents a 100% electric vehicle, without carrying the weight of the batteries
- It represents the cheaper method to electrify the transportation of goods, when compared with diesel or battery electric vehicles
- The overall savings in energy consumption, taking into account from the primary energy sources, is between 5% and 10%
- No limitation of access, as it can be operated as a standard internal combustion truck
- The existing road transport capacity can be converted into ER.

The disadvantages of ER derive from the limited access for certain areas, that makes necessary to use the fossil fuel mode of operation. Additionally, once that the truck is operated under the fossil fuel mode, the trolley mechanism becomes a dead weight.

According to this set of advantages and disadvantages of the ER, the adoption of such road transport scheme, will depend on the configuration of the transport network. That is, it can represent a practical solution for industries and business located along existing highway freight/logistic corridors.

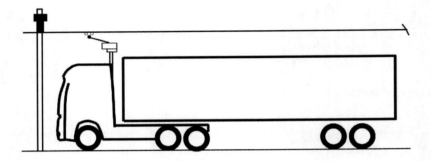

Figure 10. Electric roads schematic representation.

STUDY CASE: EUROTUNNEL

The Eurotunnel ("Le Shuttle" in French), also known as Chunnel (Channel tunnel), was an ambitious engineering project to communicate France and England through the English Channel. This channel was, until the day of the Eurotunnel inauguration, a very busy ferry-crossing space for carrying both cargo and passengers between these two countries. The Eurotunnel connects France with England along a path that is 50.5 km long, of which 37 km runs under the sea. As any other underground infrastructure, the reliability and safety demand for this tunnel is at the top level. The provisions for having the maintenance operations were designed since the very initial stages of the project, including its conceptual design, incorporating crossover points to prevent any traffic disruption due to maintenance operations. The Eurotunnel conveys nowadays, more than 36% of the cargo traffic along the strait, measured in million of trips/tons.

The current average daily tunnel traffic is 350 trains, with the truck shuttles carrying up to 32 heavy goods vehicles, each weighing up to 44 tons. The 800 meter long trains run at 140 km/h, and represent a cargo moving of 120 million tons each year, including dangerous materials which do not pay any extra charge.

The traffic along this infrastructure is thus characterized by being of diverse nature, including critical components for the industry in many economies. Any disruption of this traffic leg could thus represent big losses to all the many stockholders involved, including the operating company, the many tunnel-users, and the consigns.

Geometric Design

The overall design of this infrastructure is described in the parts of Figure 11. While part (a) of this figure illustrates a plant view of this facility; part (b) illustrates the longitudinal cross section of the Chunnel. The "tunnel" is composed of three tunnels: the north, the south and the service tunnel. According to these descriptions, the railway path does not involve short curvature turnings, nor stepped down or up – slopes. However, inside the tunnel there are transitions from the north to the south running tunnel. This means that the rail damage could be concentrated in some portions of the tunnel, corresponding to the change of track. Nevertheless, the overall damage of the railway due to the train dynamic loads, could be small in comparison with a infrastructure involving short turning radiuses and/or stepped slopes. As it was mentioned above, the turning maneuvers involve lateral load transfer that could potentially affect the integrity of the rail; and large positive slopes imply greater traction forces and, consequently, larger rail stresses leading to rolling contact fatigue. The maintenance operations involve the reinforcing/replacement of certain segments of the railway support, so that the configuration of the track allows the possibility of having some segments out of operation, while keeping the traffic alive.

Figure 12 illustrates the disposition of the railways inside the tunnel of this critical infrastructure, involving X crossings for the change of the trains from north to south running tunnels, where it can be noted the different intervals into which the infrastructure is divided (6 intervals).

There is a central tunnel that communicates at specific spots, with both running tunnels.

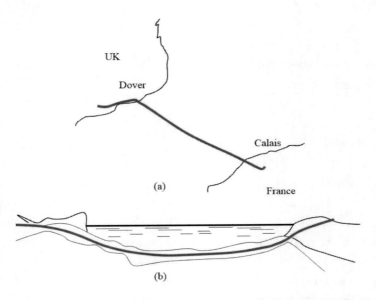

Figure 11. Schematic representation of the Eurotunnel: (a) Plant view; (b) Cross sectional view.

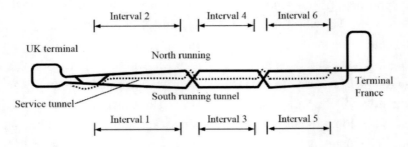

Figure 12. Railway segments of the Eurotunnel.

From the maintenance perspective of the different railway intervals, the configuration of the railways within the tunnel allows to remove from operation any of the intervals or railway segments, to perform maintenance operations, weather to the track or to the catenary. According to this railway path, there are relatively tight turns at the end of the payload segments, in order to perform the 180-degree turning maneuver.

Improving Railway Life through Instrumentation

This infrastructure owns unique preventive maintenance instrumentation to monitor the level of stress in the rails, which are continually welded and are supported by pre-cast concrete sleepers that are embedded in a concrete track bed. The instrumentation included in this infrastructure in relation with its maintenance, involves strain gages that measure the level of stress in the rail. According to the operator of the Chunnel, such instrumentation infrastructure will make possible to extend the life of the railways up to 30%. In terms of the tonnage on the rails, this represents changing the life of the rails, from 800 million to 1,300 million of tons.

The reasoning behind this remarkable rail life improvement, has to do with the maintenance operations made on the railway supports. It is well known that a failed railway support involves greater stresses on the railway (Dahlberg, 2010), and some simulation results of the level of stress as a function of a central weak sleeper, where presented above.

Monitoring the level of stresses in the rails makes thus possible to sense the condition of the railway supports, in such a way that these supports, whose cost is considerable lower with respect to the cost of the railways, are replaced in order to avoid the premature failure of the railway. In this context, specific deficiencies are being identified regarding the welding of the rail segments, as a result of damages caused during the operation of the railway switching system.

Management of the Operation of the Eurotunnel

The operation management principles of the Eurotunnel correspond to a common infrastructure management scheme, according to which a private company administrates a public asset, conforming a private-public partnership that involves many activities that are conceded to the private company, including the maintenance and the operation of the infrastructure. The principles considered for the administration of this facility, correspond to what is known as *Operations Management*, involving a series of functions, such as Strategic Planning, and Research and Engineering, amongst others. The aim of that approach, consist of providing the maximum quality to customers, optimizing resources in order to get the maximum benefit. For that, a number of surveys are performed each month in this facility, attaining up to 91.4% of satisfied/very satisfied customers in 2016, as reported by the company. Additionally, the administrating company employs "Mystery shoppers" to anonymously travel the tunnel, to provide objective measures of the quality of service.

Competitiveness and Sustainability of the Eurotunnel

In relation with the environmental impact of the railway transportation, the Eurotunnel seems to be focused on sustainability. According to the company, the system has put into service several wind turbines with a capacity of 800 kW, resulting in a total capacity of 2.4 MW. Even though this power would be not sufficient to fulfill the power requirements of the transport, which is, just for one train, on the order of 3 MW, such green energy can mitigate in a certain measure the environmental impact of such transportation system.

CONCLUSION AND RECOMMENDATIONS

Managing and accessing the railway transportation infrastructure represent critical endeavors for infrastructure operators, that should look for providing an economic and safe mode of transportation to railway users. In this context, a poor infrastructure can represent external costs to rail users, as a result of dynamic operating conditions in which the payload and railway car components can be subjected to a potentially exacerbated damage. On the other hand, delays in the transportation due to rail works, is another element posing risks to the operation and viability of this mode of transport.

Of particular interest is the maintenance of switches/turnouts, that create dangerous situations and cause damages to the mobile infrastructure, besides causing delays when such components have to be repaired while in service hours. In particular, a long reach project in Germany, involves the review of the operational principles of such mechanisms, aiming at increasing their reliability and maintainability, as any maladjustment or malfunctioning in these mechanisms, also imply higher impact loads and wearing of the rolling equipment, in particular, the wheels. Research on the damaging effects of defective and non-defective switches, on the railcar wheels, has revealed significant negative effects. However, some research still needs to be performed to analyze the potential dynamic interaction of the cargo and the carrying vehicle, which could exacerbate the negative effects of the wheel-loads on the track. The current research in this area thus could include a broader approach, in order to have a more general and realistic knowledge, on the basis of validated theoretical approaches.

Extending the life of the railway materials has been a priority for some monitoring systems reported in the literature, which are based upon sound technological principles. The advanced technological level of such components involves not only the intrinsic principles of operation used in these devices, but the whole system monitored by the system, namely, the components supporting the railways. However, the implementation of such monitoring technologies is highly dependent on the profitability of the railway network.

The Eurotunnel represents a transport system subjected to the highest demanding situations, as a result of the type and volume of traffic, as well as the physical boundary conditions that it faces. However, such infrastructure can represent an ideal situation of how to cope with the technological challenges associated to reducing the maintenance costs, while increasing the availability of the transport. Extrapolating this all-inclusive approach to other companies, could improve the competitivity of this mode of transport, although the investments made in the needed systems must be justified as a function of the potential demand.

In this chapter the challenges facing railways have been illustrated through the description of two important events occurring in different geographical areas. While the explosion in Quebec, Canada, points out the importance of taking away the transportation infrastructure from highly populated areas; the disruption of the European railway network in Rastatt, Germany, describes the importance of designing alternative routes to critical railway segments, together with the need to have contingency plans to face extraordinary situations regarding the connectivity in a given transport network.

Future research efforts are recognized in relation with the railway car – track interaction, as the dynamic loads derived from such interaction could be reducing the life of such infrastructure. It is particularly important for infrastructures dedicated to the transportation of liquid hazardous substances.

REFERENCES

Berkeley, T. (2017). Rastatt was a disaster for rail freight. Let's learn the lessons, *International Railway Journal*, October 2, 2017.

Boschert, S., Schulze, M., and Salson, L. (2017). Railway Switch Point Machine simulator (RaSPoMaS) for simulation driven operation and maintenance, *Proceedings, NAFEM World Congress 2017*, Stockholm, Sweden, 11-14 June, 2017.

Connolly, D. (2017). Economic viability of electric roads compared to oil and batteries for all forms of road transport, *Energy Strategy Reviews*, 18, 235-249.

Dahlberg, T. (2010). Railway Track Stiffness Variations-Consequences and Countermeasures, *International Journal of Civil Engineering*, 8(1), 1-12.

Doulgerakis, E. (2013). *Influence of switches and crossings on Wheel wear of a freight vehicle*. Master of Science Thesis, Royal Institute of Technology, Sweden.

Gaudry, M., Lapeyre, B., and Quinet, É. (2016). Infrastructure maintenance, regeneration and service quality economics: A rail example, *Transportation Research part B*, 86, 181-210.

Nielsen, J. C. O., Pälsson, B. A., and Torstensson, P. T. (2016). Switch panel design based on simulation of accumulated rail damage in a railway turnout, *Wear*, 366-367, 241-248.

Rodrigues Vieira, D., and Lavorato-Loures, P. (2016). Maintenance, Repair and Overhaul (MRO) Fundamentals and Strategies: An Aeronautical Industry Overview, *International Journal of Computer Applications*, 135(12), 21-29.

Romero, J. A., Alvarado Morales, M. A., Arroyo Contreras, G. M. and Betanzo-Quezada, E. (2016). *Modelling of Dynamic Bending Stresses due to a Weakened Rail Support*. Internal report. Queretaro Autonomous University.

Stenström, Ch., Norrbin, P., Partida, A., and Kumar, U. (2016). Preventive and corrective maintenance - cost comparison and cost-benefit analysis, *Structure and Infrastructure Engineering*, 12(5), 603-617.

Xin, L., Markine, V. L., and Shevtsov, I. Y. (2016). Numerical procedure for fatigue life prediction for railway turnout crossings using explicit finite element approach, *Wear*, 366-367, 167-179.

In: Transportation Infrastructure ISBN: 978-1-53614-059-0
Editor: S. Antonio Obregón Biosca © 2018 Nova Science Publishers, Inc.

Chapter 6

URBAN POLICIES FOR SUSTAINABLE MOBILITY: THE RETURN OF THE TRAM TO BARCELONA

Pere Macias[1,2,#], *Antonio Gonzalez*[3], *Abel Ortego*[4],
Elisabet Roca[5], *Robert Vergés*[2], *Josep Mercadé*[2],
Joan Moreno[6], *Alessandro Scarnato*[7],
Ana María Moreno[8], *Jesús Arcos*[9], *Clement Guibert*[10],
Marianna Faver[11], *Etienne Lhomet*[12]
and Míriam Villares[5,*]

[1]Barcelona City Council, Barcelona, Spain
[2]Barcelona School of Civil Engineering, Universitat Politècnica de Catalunya, Barcelona, Spain
[3]Urban Planning Agency (A'urba), Bordeaux, France
[4]CIRCE Foundation, Zaragoza, Spain
[5]Institute of Sustainability Science and Technology, Universitat Politècnica de Catalunya, Barcelona, Spain

---

[#] Strategic Director of Tram Connexion Project
[*] Corresponding Author Email: miriam.villares@upc.edu.

⁶Vallès School of Architecture, Universitat Politècnica de Catalunya, Barcelona, Spain
⁷Universitat Politècnica de Catalunya, Architect and Historian, Barcelona, Spain
⁸Zaragoza Tramways, Zaragoza, Spain
⁹Aldayjover, Architecture and Landscape, Barcelona, Spain
¹⁰Light Rail Transdev-Australasia, New South Wales, Australia
¹¹Deltametropool, Rotterdam
¹²Des Villes et Des Hommes, Regional Planning and Mobility, Bourdeux, France

Abstract

In the last 30 years, many European cities have opted for the tram as a quality public transport mode for cities. Unused lines have been renovated and new lines and networks have been created. The success of new tramways comes from their efficiency as a transportation mode but also from their restructuring of the urban fabric and transformation of public space. A particularly interesting case is Barcelona, which not only has reintroduced a tram system (eliminated in the 1970s) so as to link the city with its suburbs, but now has plans to reconquer the city centre in a strategy aimed at sustainably improving the efficiency of the city's public transport network at the metropolitan scale.

Keywords: tramway, sustainability, urban project, transportation systems

Introduction: Three Reasons for the Return of the Tram[1]

The tram has survived in many countries of the world and is being modernized. In Europe, Berlin, Vienna, Amsterdam, Brussels, Helsinki, Lisbon, Turin and Zurich have preserved their networks and are

---

[1] In this section Antonio González, Pere Macias, Abel Ortego, Elisabet Roca and Míriam Villares have contributed.

developing and updating them; Geneva, Porto and Valencia have reactivated lines; and Manchester, London, Sheffield, Nantes, Grenoble, Lyon, Paris and Bordeaux have rebuilt lines.

In Barcelona, the tram, which had disappeared in 1971, returned with the Baix Llobregat and Besòs tramway projects (Trambaix and Trambesós), inaugurated in 2004. These two lines, however, connect the city with its suburbs, whereas linkage via the central section of Avenida Diagonal is a logical aspiration from the point of view of the topology of the public transport network (see Figure 1). However, this linkage has not happened for reasons such as opposition from local business associations, urban difficulties, cost (against the background of the financial crisis that started in 2008), fears of negative repercussions on traffic, etc. People who oppose the project resort to the main argument that the tram is not a suitable transport medium for a city as densely populated as Barcelona, insisting that the underground is a more suitable solution for a city with so many activities and flows in its public spaces. Now that the municipal government has become involved in the debate, it seems appropriate to recall that one of the great advantages of a modern tram way system is its great versatility as a tool of urban planning, especially in highly polluted environments.

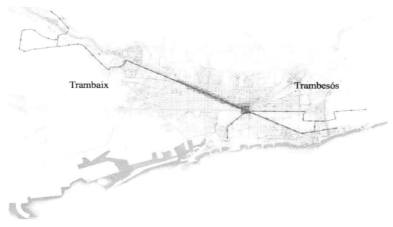

Source: Barcelona City Council.

Figure 1. Trambaix and Trambesós in Barcelona.

Modern tram systems are successful for three main reasons. First, they are an efficient, reliable and comfortable transport mode. Second, they allow better structuring of urban projects than other means of public transport. Third, more than other transport mode, they invite reflection regarding the distribution of public spaces, the role of the private vehicle and environmental and public health issues.

An Efficient and Attractive Mode of Transport

As a mode of transport, the modern tram has several advantages over other urban overground public transport modes, mainly buses. It can transport double or triple the passengers of an articulated bus.[2] It can usually travel faster; the commercial speed depends on how it is adapted to the urban context, but a completely segregated modern tram can reach 20 km/hour, while bus lines rarely exceed 15 km/hour in the best of cases; segregated circulation also ensures great regularity, allowing networks to operate at 3-minute intervals during rush hour. Finally, trams are very reliable, which makes them popular with users and also draws users from other less reliable public transport modes.

The best evidence of the success of trams is the number of passengers using them in the densely populated cities in which new networks have been implemented.

It can be argued that segregated bus lanes with traffic light prioritization would achieve high bus speeds and high levels of reliability and regularity. This is certainly true, but there are several obstacles preventing the bus from taking on the same role as the tram. Its capacity does not allow it to transport the same number of passengers/hour, and, although the fact that it does not need a particular infrastructure is an advantage (lower cost and greater route flexibility), this is also a drawback. Bus lanes may be interrupted if the available space is scarce or they may have to be available to facilitate traffic turns to the right or left.

---

[2] This varies depending on the model, but the articulated buses usually used in European cities have a capacity for about 120 passengers, while that of the tram is about 300 passengers.

Furthermore, since this space may be shared, the risk of accidents is aggravated. Segregation of the tram, however, is fixed and exclusive and so has to be respected along the entire route. Finally, the tram also provides greater comfort to passengers in the form of brighter and wider spaces, less noise, smoother movement and better accessibility for people with reduced mobility.

It could also be argued that the underground offers greater capacity and still better operating conditions in terms of guaranteeing reliability and regularity. But, among other drawbacks,[3] the underground lacks one characteristic of the tram that explains much of its success: its ability to transform cities.

An Urban Planning Instrument

The examples are many and before and after photographs are spectacular: modern tram projects in Europe are usually accompanied by a complete overhaul along tramways from beginning to end. Public space is recovered for pedestrians and cyclists, for trees and plants, for seating, public fountains and other amenities. Crossroads and pedestrian crossings are redesigned, pavement accessibility is improved and traffic is slowed down. Tram networks are the perfect instrument for designing new urban layouts (see Figure 2), most especially in very built-up areas; one example is historic centres with narrow streets, where the need for cars to circulate and park significantly reduced pavement space. Tram infrastructure requires major reforms of roadways, pavements and sometimes even the subsoil that displace service networks and so are usually implemented with urban improvement in mind.

In addition to reorganizing the public space, new tram lines crossing city centres usually accompany, in a broader sense, urban projects such as the rehabilitation of old buildings and the construction of new amenities. Building sports or cultural facilities near the new line is a way to attract

---

[3] Its cost, its lower accessibility and the fact that it usually circulates underground, which gives it to more claustrophobic character.

new users to public transport. The private sector also views the new infrastructure as revaluing land and property through improvements in accessibility and environmental quality.

Source: Agence d'Urbanisme Bordeaux Métropole Aquitaine.

Figure 2. Urban and landscape integration of the tramway in Bordeaux.

Finally, in the less dense, suburban areas, tramways allow the territory to be planned and organized along new public transport axes, with layouts that aim to connect out-of-town strategic sites (universities and business areas) and outlying housing estates with city centres. The territories crossed by the tram are affected in a positive way and the visibility of the tram infrastructure becomes one of its main virtues, given that it becomes a feature of the territory, indicates the direction of the centre and represents a link with the rest of the city.

Environmentally Friendly Transport

Despite its use of electrical power might come from non-renewable sources, the truth is that the tram is perceived to be a clean mode of locomotion. It consumes ten times less energy and generates a hundred times less greenhouse gas than an automobile (CERTU, 2002). It also causes less noise pollution than buses. The ecological image of the tram is also sustained by the accompanying new infrastructures, typically involving new green areas, including grassed tramways, which reduce the visual impact of the infrastructure and is particularly protective of parks and gardens.

However, the tram is a powerful instrument for environmental renewal mainly because it reduces the space available to cars. The widening of pavements and the creation of bike lanes and even bus lanes tend to be done without any great questioning of the place of the car: bicycles and pedestrians do not need much space and bus lanes can be used by cars in certain circumstances. Since the tram, however, has to be segregated to a greater or lesser degree along most of its route, the space for cars has to be reduced, bringing with it numerous environmental and public health benefits.

Undoubtedly, a tram network is not required to reorder the public space, plant trees, reduce the space for cars or improve the environmental quality of our cities. But a tramway project does help make political decisions in a specific direction. On the one hand, the general interest usually neutralizes the arguments of opponents, and on the other, politicians can act, decide and show the benefits of their actions (Offner, 2001). Indeed, decisions that lead to a change in the habits of citizens are more easily made if structured around a project such as a new tram line or system.

Cities host very varied activities and mobility for their inhabitants is increasingly complex, such that no single mode of transport can meet all travel needs. The best solution for mobility is to offer a range of possibilities, while taking into account environmental quality criteria. The tram is one of the better options because of its carrying capacity, its relative environmental friendliness (clean and silent) and its traffic calming qualities. Nonetheless, it has to be coordinated with other transport modes (as well as the car), including the underground, bus, bicycles and pedestrians. The issue, ultimately, is to create a varied and efficient citywide transport system.

The introduction of a tram system in a city changes the mobility patterns of inhabitants. The three main changes are as follows:

- Road traffic. Implementation of a tram system reduces circulation and parking space for private vehicles and also has a calming effect on traffic. Traffic light prioritization and platform systems ensure that trams can achieve speeds of 20 km/h (commercial speed) — often higher than the average speeds achieved by private vehicles.
- Other public transport modes. Implementation of a tram system as a high-capacity means of transport leads to restructuring of other public transport modes, but especially buses, and so offers public transport users a less polluting alternative.
- Suburban connectivity. Good tram system design can act as a backbone for a city by offering suburban dwellers — who usually depend more on private vehicles — cost- and time-effective and efficient transport to the city centre.

Another consequence is improved air quality, as there is greater use of non-polluting means of transport in response to urban space newly made available to cyclists and pedestrians. The outcome of the overall modification to the mobility patterns of a city is reduced greenhouse gas emissions, mainly of carbon monoxide (CO) and nitrogen oxide (NOx), but most especially of solid particles (PM2.5 and PM10), which are a special health risk according to the World Health Organization (WHO).[4]

Yet another consequence is energy savings. The main source of energy savings is the occupation factor. Good tram system design therefore ensures a high passenger-kilometre index (PKI) — more than 20, compared to that of an urban bus, which rarely exceeds 5[5] A tram is more efficient because its electric traction performs better than a conventional

---

[4] WHO report at http://www.who.int/en/news-room/detail/27-09-2016-who-releases-country-estimates-on-air-pollution-exposure-and-health-impact.

[5] See the Metropolitan Mobility Observatory report (in Spanish) at http://www.observatoriomovilidad.es/images/stories/05_informes/Informe_OMM_2015.pdf

combustion engine as used in urban buses. Figure 3, which exemplifies this measure for the city of Zaragoza, compares energy consumption per unit of distance and per unit of distance-passenger transported. Thus, energy consumption per unit of distance for bus and tram is very similar, at 5.9 kWh/km and 3.9 kWh/km, respectively. Nevertheless, when indexed to average passenger occupation, the Figure of 1.2 kWh/passenger*km for buses drops to 0.18 kWh/passenger*km for trams, reflecting 85% lower energy consumption for trams compared to buses.

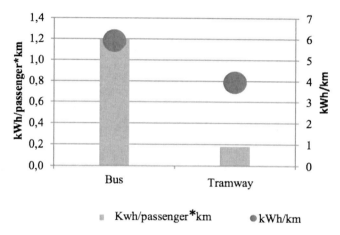

Source: Z2020xMUS Project, 2014.

Figure 3. Energy consumption for Zaragosa buses and trams compared.

A tram system thus enables local objectives to be achieved in the fight against climate change,[6] typically expressed in terms of reduced CO_2 emissions. Considering both direct and indirect emissions caused by the generation of electricity, global emissions for a tram are 71.49 $grCO_2$/passenger*km compared to 328 $grCO_2$/passenger*km for a bus.

Nonetheless, the mere fact of installing a tram system does not guarantee success, unless it is well designed, meets the mobility needs of

---

[6] These objectives are usually reflected in voluntary agreements such as that of the European Commission referring to the Covenant of Mayors programme, whereby cities commit to reducing their CO_2 emissions by at least 20% by 2020.

citizens and takes into account areas lacking in effective public transport. Examples of Spanish cities where the impact of a tram has been less than expected are Malaga, Seville, Parla and Granada (Carmona, 2015). Evaluation ex-ante of the impact of a tram system is a complex task that requires much planning. The proper approach is not just to implement a new means of transport, but also to satisfy the real needs of a city's inhabitants and ensure that it is the best possible mobility option for a particular city. The following issues in particular need to be considered:

- Optimal design that ensures adequate coverage of areas of the city that are attraction centres (hospitals, leisure areas, universities, city and neighbourhood centres).
- Systems that ensure a commercial speed superior to that of other transport systems (traffic light prioritization, exclusive platforms, etc).
- Restricted space for private vehicles (pedestrianization, fewer lanes, urban tolls, fewer parking areas).
- Service frequencies adapted to mobility needs: high frequencies at peak hours and medium-low frequencies at times of less demand.

The experiences of several cities discussed below highlight different aspects and reasons for the success of the modern tram.

CASE STUDIES

Introduction

On a global scale, around 400 cities have incorporated trams in their public transport networks, of which around 200 are cities in Europe.

Incorporation of the tram has motivated changes that vary depending on the geographical model. Here we focus on the French and German models. The French model reintroduces the tram, re-urbanizes areas and renews neighbourhoods on the basis of new, much more efficient and modern transport lines that improve quality of life for the inhabitants of the areas crossed by the tram. Bordeaux is a good example of this model, and also Rome, where the tram was introduced in the historic centre. The German model is featured by sound political decision-making that, in the past, decided not to abolish trams networks. This factor helped renovation of such networks as the basis for city transport systems. Some Central Europe and Balkan networks reflect this model, although they have not achieved the same level of modernization for economic reasons.

The tram system in Barcelona — considered an example on an international scale whose developments are followed by many cities — has a central stretch of Avenida Diagonal that remains incomplete and which is perceived as a very important issue for the inhabitants of the city. Tram system implementation in other cities are being observed with a view to approaching the problem of the unbuilt stretch. However, each city has its own particular morphology and so has to seek its own solution to implementing this mode of transport. Despite differences, nonetheless, similar situations can be used as a reference in facing the challenges and risks associated with implementing tram systems and, in Barcelona, connecting up a tram system. Some questions arising from the challenge of implementing a tram system are related to nostalgia for the past and commitment to the future, connectivity and compatibility with other transport modes, effects on gentrification, the most suitable management model, etc. Below we describes tram systems in Lyon and Bordeaux, Rotterdam and Amsterdam, Zaragoza, Sydney and Florence.

The Tram and Metropolitan Mobility in France: Lyon and Bordeaux[7]

Description

Bordeaux opened its network of three tram lines with an average speed of 18.2 km/h in 2003. Implementation was initially complicated by

---

[7] In this section Ethienne Lhomet and Antonio González have contributed. The complete case of study can be found in: https://wwwtramvia.webs.upc.edu/index.php/casos-de-estudio/.

relatively sparsely populated areas where it was difficult to justify the presence of a tram system.

Table 1. Lyon and Bordeaux tram networks

	Bordeaux	Lyon
Organizing authority	Bordeaux Métropole	SYSTRAL
Operator	Keolis	Keolis
Year of commissioning	2003	2001
Length	3 lines (+1 in construction)	6 lines (counting the Rhône Express connection with airport)
	66Km (77km with line D. planned opening in 2019)	66Km (85km with new lines planned)
Stops	116 stops	107 stops
Number and capacity of trams	105 tramways (+15 in order)	91 tramways
	Trams of 33 and 44 meters (maximum capacity 300 persons)	Trams of 33 and 44 meters (maximum capacity 300 persons)
Maximum speed	60 km/h	70 km/h
Commercial speed	18.2 km/h	22.0 km/h
Maximum frequency (in HP)	One tram every 3 minutes	One tram every 3 minutes
Evolution in number of passengers	2009: 59.4M	2010: 50.5 M
	2011: 66.5 M	2014: 82.4 M
	2016: 80.0 M	2015: 85.6 M
	+35% in 7 years (or +20% in the last 5 years)	+43% in 4 years

Source: Agence d'urbanisme Bordeaux Métropole Aquitaine.

Lyon — a densely populated city with around 10,000 people per square kilometre — opened its tram network in 2001 and currently has five tram lines with an average commercial speed of 22 km/h (largely due to its faster suburban tram-train line). In 1974, Lyon relied on private cars and public buses. A network of four underground lines was built, but did not help reduce the use of private vehicles. The construction of new lines was halted and investment was redirected to the implementation of a tram network and new trolleybus lines. This network is very integrated with the underground and 74% of journeys are made by trolleybus, underground or tram, all electrically powered. The objective of the tram system in Bordeaux— understood as a tool for implementing urban projects, ensuring fast and comfortable displacement and improving urban quality of life — is to promote multimodality, reorganize road systems and expand public space for pedestrians (see Table 1).

Trams systems are viewed as an instrument for urbanization as they lead to quality urban design and architectural refurbishment accompanied by the reintroduction of nature in the form of green spaces, thereby enhancing territorial cohesion, structuring the urban fabric and acting as a vector of social integration. A good outcome requires a thorough understanding of people's mobility needs, but a transport project cannot succeed unless it is underpinned by an urban vision and supported by efficient political leaders, banks, local businesses, among others.

Impact

The tram network of Bordeaux, with 800,000 inhabitants, has some 80 million passengers per year and the annual increase in use is around 12%. In Lyon, most public transport use is focused around the tram.

Implementation of new tram networks in France has led to a decrease in the use of cars in French cities. The goal is to reduce displacement by car by 60% by promoting the development and use of collective transport by 25% and of the bicycle by 15%.

The problem of justifying tram line implementation in sparsely populated areas was solved through programmes to coordinate urban planning with mobility projects and through agreements to densify cities, which also required public land management mechanisms to be put in place.

The implementation of tram systems has led to a reduction in operating deficits due to the considerable increase in fare revenues. Enlargement of networks does not imply an increase in deficits, as this benefits from the increase of the number of passengers. Tram systems are thus a source of revenue that can be invested in other urban areas, activities and actions.

Lyon used its tram project to transform the city's historic centre into an outdoor shopping centre, thereby improving mobility of its citizens and at the same time stimulating commerce. The tram project led to an increase of 10% in the number of shops and this trend has been maintained since the tram was inaugurated, with 42% of shopkeepers reporting an increase in business in the period from announcement of the tram project.

Table 2. Decrease in the use of cars

	2009 (veh/day)	2013 (veh/day)	2013 corrected (veh/day)	Δ tram (%)
Traffic in first belt	70,410	60,285	65,481	-7.94
Traffic in second belt	351,950	307,088	327,314	-6.18
Traffic in third belt	412,690	389,249	383,802	1.41
North direction	133,290	75,845	123,960	-38.81
South direction	137,330	75,804	127,717	-40.64
Both directions	270,620	151,649	251,677	-39.74
Total City	2,159,142	1,853,720	2,008,002	-7.68

Source: Agence d'urbanisme Bordeaux Métropole Aquitaine.

The Tram as the Lynchpin of Public Transport in the Netherlands: Rotterdam and Amsterdam[8]

Description

The Randstad, located on a large delta, is a connected polycentric system of functionally specialized cities (Amsterdam, Rotterdam, The Hague and Utrecht), with some 7 million inhabitants. Nearly 50% of the Dutch population lives in this area, occupying a mere 20% of the country's land area, and 45% of the country's GDP is produced here. At the urban planning level, the Randstad as such did not exist until the late 1960s.

In 1998 the mayors of the Randstad's main cities signed a manifesto to update the Randstad to what would be called the Deltametropolis. This renewal would be based on integral planning of mobility around traditional centres of activity. It was agreed to create a transport network that would

---

[8] In this section Joan Moreno and Marianna Faver have contributed. The complete case of study can be found in: https://wwwtramvia.webs.upc.edu/index.php/casos-de-estudio/.

newly structure this entire territory on the basis of transport nodes, intermodal stations and the large central stations of the four cities (see Figure 4). The Randstad currently has 126 railway stations, 95% and 100% of which have carparks and bicycle parks, and 22% and 82% of which have modal interchange points with tram lines and bus lines, respectively. The transport system is managed by different companies that very productively coordinate their activities.

The two fundamental stages in urban planning in the Netherlands are urban development and the transport network. Urban planning oriented to transport concentrates urban development around transport networks. Five elements are considered to ensure an attractive mobility system, namely, service diversity, short distances and the quality of stations, the space and the design.

The Deltametropolis Association has a number of projects in the pipeline, including the Sprint City project, a game that simulates a real city that has to be equipped with an efficient mobility plan, and the Maak Plaats book developing the butterfly model based on distinctive features of both the network and the physical space.

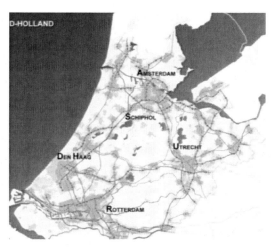

Service influence
Motorways network
Access to the motorway network
Railway network
Access to rail network

Source: J. Moreno (2017).

Figure 4. The Randstad tram-train and road network and service influence.

RandstadRail is an integrated public rapid-transit system composed of tram, bus and underground lines that link up Rotterdam and The Hague central stations and Rotterdam underground.

Impact

The impact of the transport network on the compact model of densification makes transport more efficient and establishes a hierarchy around the city of Amsterdam. The intermodal hierarchy develops from an international node, as happens with the Zuidas project, or from an urban node, as happens with the Beimer project. There is also the possibility of a third node as a space of exchange at the local level.

It is interesting that where there is a tram, there is a city and where there is no tram there is no city, i.e., the tram network reflects urban structure. There is, therefore, a direct correspondence between the shape of the city and the shape of the network, with the tram as the element that orders, even physically, new developments. Urban development, transport mobility and local activities must be designed together as one is not the consequence of any other. An example is developments and density increases around stations.

Also operating in the Netherlands is the OV-chipkaart as a payment method that works in the entire public transport network of the country and that may even possibly be extended to Brussels.

The Tram as a Key Element in Environmental and Urban Renewal: Zaragoza[9]

Description

Zaragoza is a dense city of 700,000 inhabitants that acts as a geostrategic hub in Spain and southern Europe. The existing tram line follows a north-south route through the centre that crosses neuralgic points

---

[9] In this section Jesus Arcos, Ana María Moreno, and Abel Ortego have contributed. The complete case of study can be found in: https://wwwtramvia.webs.upc.edu/index.php/casos-de-estudio/.

of the city, including business areas and two social housing districts at each end. The tramway acts as the central axis of the city and unites concentric areas.

The line is 12.8 kilometres long, has 25 stops and runs 21 trams, 18 trams with a capacity for 200 people at 5-minute intervals during peak hours. Power supply is catenary except in 1.8 kilometres of the central zone (the longest non-catenary distance in Spain). The average distance between stops is 500 metres. There are three interchange connections with buses and suburban train and two intermodal subsidized carparks (€0.06 cents per hour if the tram is used). Fares are integrated with urban and interurban bus fares.

The technology used is environmentally friendly and energy efficient. The trams have fast-charge accumulators that save up to 35% of energy. Canopies designed for the tram was provide shade that benefits from thermal inertia with the help of sedum plants. Specially designed cabinets contain all the necessary facilities at every tram stop, ensuring a tidy public space.

The tramway, has an average of 100,000 passengers on weekdays, which makes it the most used such transport in Spain. Demand in 2016 was 28 million and a total of 124 million passengers have been transported to date. Use of the tram is 30% of bus user, yet the tram only covers 7% of the kilometres covered by buses.

The tram is run by a mixed company with Zaragosa City Council holding a 20% share and with 80% in private hands (CAF, TUZSA, FCC, ACCIONA, IBERCAJA and CONCESSIA).

Impact

The tram has proven to have a great social return — one of the most important factors in these kind of projects — with a fraud rate of around 1%. It led to a renewal of streets and avenues and has reinvigorated businesses. The tram operates on the surface and had a showcase effect that has even motivated the publication of a shopping magazine called *Tranvida*.

Around 24% of users have stated that they have replaced the car with the tram and 52% of users that they use the tram daily. Nearly 16 million passengers have used the tram instead of the bus. In 2016 there were 27.9 million uses of the tram and 80.6 million uses of the bus. More important is the fact that the tram has 20.34 users/km, compared to 5.04 users/km for the bus.

The greatest impact has been the reduction in traffic and, consequently, in energy consumption (at 189,661.64 megawatts/year, this means that almost 20 million litres of fuel is saved per year). Traffic to the centre has been reduced by 32% and elsewhere by 50%. Traffic accidents have been reduced by half over the last decade. The dynamic traffic light prioritization system also allows an estimated saving of 8% of the energy used by the trams in that it is not necessary to brake and pull.

Furthermore, emissions have dropped by 15%. The resulting improvement in air quality has led to a fall in emissions of CO, NOx and PM10 by 11.7%, 9.45% and 47.8%, respectively, in the downtown area of Zaragoza. Figure 5 compares PM10 emissions in the centre of Zaragoza before and after the tram was inaugurated. The Zaragoza tram was the first tram in the world to obtain the ISO 14025 certificate for its contribution to protecting the environment.

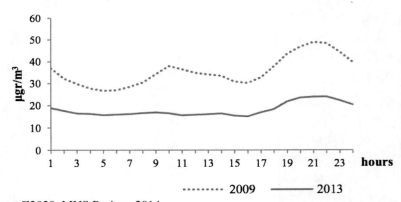

Source: Z2020xMUS Project, 2014.

Figure 5. Hourly PM10 emissions in the centre of Zaragoza.

In terms of greening, 1,200 trees were planted (in 2016, 113 trees). A total of 42,000 m² of green area has been planted, including the sedum plants on tram stop canopies.

Finally, accessibility for people with reduced mobility is one of the factors in the success of the tram, which has been recognized with awards from ONCE and the Foundation DFA.

The Tram at the Heart of the City: Sydney[10]

Description

Sydney, the capital of New South Wales, covers a very large area (1,687 km²) and is featured by dispersed residential suburbs and very few city highways. It is the most populated city in Australia, as its metropolitan area has 4.34 million inhabitants. Transport in the Sydney area is based on roads and a railway that is quite widespread in the territory. An important network of commuter railways has been established that take advantage of existing railway lines.

In 1988, the monorail came into service in the city centre, although its connectivity with other transport networks was initially low. For a long time it was thought that the monorail would be the transport of the future, but it was decided that this was not the case for Sydney and so it was closed.

A disused goods transport line was partially used to build a tram line that was opened in 1997 and paid for by a public-private partnership contract. The line is 12.8 kilometres long, has 23 stations and uses Urbos CAF rolling stock, made up of 12 units (see Figure 6).

A second tram line is currently being built to connect the Central Business District and the south-east. The new line is 12 kilometres long, has two branches, 19 stations and a 400-metre long tunnel. It also has a section of 1.5 kilometres in length without a catenary to avoid problems of visual impact. The capacity of the line is approximately 9,000 passengers

---

[10] In this section Clement Guibert and Robert Verges have contributed. The complete case of study can be found in: https://wwwtramvia.webs.upc.edu/index.php/casos-de-estudio/.

per hour and direction and the frequency interval is 4 minutes for the Central Business District. The trams are from the new Alstom X05 Citadis range. The line, planned to be in service by March 2019, is called the Central Business District and South East Light Rail (CSELR). According to the environmental impact study, it is divided into five sections: City Centre, Surry Hills, Moore Park, Kensington & Kingsford and Randwick.

Transport for New South Wales is in active management, planning and construction phases and design, construction, maintenance and operation are based on public-private partnership agreements, with Acciona Infraestructure Australia, Alstom, Capell Capital and Transdev as the main contractors in the Connecting Sydney consortium.

Source: Salmeron, 2017.

Figure 6. Tramway integration in Sydney.

A part underground-part aerial underground line is also being constructed, at the substantial cost of 12,000 million euros, with a territorial scope — 66 kilometres long — much greater than that of the tram.

Impact

Transport in New South Wales was originally expected to meet the needs of a million people in 20 years, but these numbers were underestimated. The two tram lines cover the central part of the city of Sydney. The first tram (or light-rail) line is currently in service. Maintenance requires 119 people and demand since 2011 has grown by 34%. The line opened in the southern area of the city, which rapidly began to densify. The line works very well and there are no problems with traffic or with the population. Regarding the second line, Randwick will have new tram parking facilities where minor maintenance of rolling stock will be carried out. Major maintenance operations are carried out in Lilyfield, where the other line's facilities are currently located.

From Tram Rejection to Success: Florence[11]

Description

Florence is a small flat area surrounded by mountains, where urban development is monodirectional towards the west and northwest and connected with the entire metropolitan area, including Pistoia and Prato, with more than 1.5 million people.

The tram arrived in 1871 — through the Société Générale des Chemins de Fer Économiques — as a solution to connecting different levels of the city. This modern and popular system interfered with visual perception of the city and was marked by frequent accidents.

In the 1950s, the tram network was about 200 kilometres long. The filobus appeared after the Second World War due to the safety problems with the tram, but only lasted three years due to its ineffectiveness. The war, however, had damaged the tram system, making rebuilding difficult and leading to heavy reliance on bus lines. By the end of the 1930s, with the arrival of the Fiat car factory, the car began to be promoted as an element of the urban landscape. This fact, plus the closure of the tram and the filobus, supported the use of private vehicles. Florence was the fourth Italian city in terms of vehicles per capita by the end of the 1980s, causing parking problem in a city without restrictions on vehicles.

In 1973, a Ministry of Cultural Assets was created to protect the country's heritage. In 1986, a programme to progressively pedestrianize the centre of Florence was launched and, in 1988, a reduced traffic zone was created that divides the centre into four areas; it was designed to

---

[11] In this section Josep Mercadé and Alessandro Scarnato have contributed. The whole case of study can be found in: https://wwwtramvia.webs.upc.edu/index.php/casos-de-estudio/.

prohibit access and movement in downtown areas at specific times/days other than for residents or workers.

The Marcello Vittorini Plan (1990) encouraged political discussion among citizens about the mobility model of their city, whether based on an underground or tram system. The result was approval of the tram project, with three lines that link the centre with the main points of attraction.

Romano Prodi (1998) started the construction of the public transport network, with proposals for a tram system and implementation of a high-speed train to connect Milan and Rome with the centre of the network in Florence.

Table 3. Results of 2007 Tramway referendum

	Yes (I want to stop the project)	No (I want the project to proceed)	Voters
Question 1 – Line 3 Careggi - Europa	51.87%	48.13%	39.35%
Question 2 – Line 2 Peretola - Liberta	53.84%	46.16%	39.35%

Source: A. Miglietti (2008).

Scientific studies, however, warned of the vulnerability of historical monuments to the tram. This situation led to political and social discussions regarding each part of the tram route. In 2007 a referendum was held to ascertain support for the second and third tram lines (see Table 2).

Impact

The first line has achieved an average flow of about 13 million passengers, almost double the estimate. In 2012 and 2014, the region of Tuscany carried out surveys on use of the tram, finding that approximately 25% of tram users were previously users of private vehicles. Another survey conducted by GEST in 2011 found that 54% of tram passengers had stopped using their car. The latest evaluations by local administrations in Florence refer to a reduction of 20% in traffic.

According to Cresme Consultants, by the time the three lines of the tram system are in operation, the value of property will have increased by 5-10%, on the basis of the impact observed in other Italian and European cities.

A recent survey of citizens found that 75% of a sample of 15,000 inhabitants of Florence are very satisfied with the tram, which would indicate that the implementation has been a success.

Nonetheless, several factors remain to be addressed. Integration between the tram system and the railway system has yet to be determined, both in relation to the high-speed and conventional train networks. A critical element is the city's structural plan in terms of how this could help shape the habitability and quality of the spatial configuration of the city.

DISCUSSION[12]

The success of the modern tram can be attributed to its role as an efficient and friendly mean of transport, its ability to structure urban projects and its contribution to the debate on how the distribution of urban space benefits the environment and public health.

Both the initial theoretical contributions and the case studies for several cities described above contribute to our observations regarding the case of Barcelona.

Tram Connection via Avenida Diagonal

At present Barcelona has two independent tramways (Trambaix and Trambesós), launched in 2004, with lengths of 15.4 and 14.1 kilometres, respectively, operated under a license regime and serving nine central municipalities of the metropolitan area.

---

[12] In this section Antonio González, Pere Macias, Abel Ortego, Elisabet Roca and Míriam Villares have contributed.

Demand exceeds forecasts made at the time of tender, despite the severe economic crisis that affected the Spanish economy and that led to a stagnation in the number of passenger journeys between 2008 and 2013. Recovery in mobility since 2013 is reflected in the strong growth in demand for the two lines. Trambaix transported 17.6 million and Trambesós 9.1 million passengers in 2016. Growth rates in use of Trambaix and Trambesós are currently around 4% and 8.5%, respectively — higher than the rates for the transport system as a whole.

The Metropolitan Transport Authority Consortium (ATM) and Barcelona City Council are currently processing an informative study regarding the connection of both networks through the central stretch of Avenida Diagonal. This connection involves laying 3.9 kilometres of track between Plaza Francesc Macià (origin of Trambaix) and Plaza Glòries Catalanes (origin of Trambesós). The commissioning of this section reflects a very strong increase in demand, estimated at 227,000 users per working day for 2022 (after a three-year ramp-up), double the current 93,000 users.

However, implementation of this project is not exempt from controversy. Intervention in the central section of one of the most emblematic avenues of Barcelona led to a citizen consultation in 2010, in which 12% of the municipal census participated. The municipal proposal was rejected by 80% of votes against the two alternatives put forward.

The opponents have different reasons to justify their convictions, including an aggression against the monumental character of the avenue, disruption to road traffic and disagreement with the cost of the new system. Some argued that an electric bus service would be sufficient to meet transportation needs.

As almost always happens, an alliance was established between right and centre-right sectors with trade and lobbies linked to the automobile sector. Meanwhile, the Catalan government, the ATM itself, left and centre-left parties, neighbourhood organizations, environmentalists and public transport users defend the connection as a key element in enhancing the efficiency of the integrated public transport system and as an instrument for improving air quality (a severe environmental problem in Barcelona).

The Tram as Transport Mode

In light of the cases analysed, the role of tram systems in collective public transport in cities can be contextualized in two ways: in medium-sized cities and in large cities.

The first context refers to medium-sized cities like Bordeaux and Zaragoza, in which new tram networks play a basic structuring role. The tram serves as the central element that communicates the centre with the suburbs, shaping not only transport policies but also the urban regeneration of neighbourhoods and the rehabilitation of public spaces. Urban bus routes complement the tram network as the result of integrated fare systems, coordination of flows between the respective networks and the creation of modal interchange stations. In Bordeaux and Zaragoza, the results of the implementation of new transport systems based on tramways have been spectacular, with large increases in mobility, more rational bus networks and a maximally effective urban transport model. This success of the tram in medium-sized cities has been frequently denigrated by its detractors with the simplistic argument that "cities that do not have or cannot have an underground have to have a tram system," with the corollary that a tram system is not considered necessary for cities that already have a metropolitan rail system.

The second context refers to large cities like Barcelona, Paris, Berlin, Sydney and Amsterdam, where trams do not play a major structuring role, given the existence of high-capacity commuter rail and underground systems. However, the role played by the tram in these large metropolitan areas is that of enhancing the efficiency of overground public transport (see Figure 7). As in medium-sized cities, the tram can connect the centre with the suburbs, by occupying old roads or reusing disused railway lines. Trambaix in Barcelona, for instance, led to redevelopment of the old N-

340 and C-246 passing through Esplugues and Cornellá de Llobregat. Paris has also used this formula for its T-1, T-2, T-4, T-5, T-7 and T-8 lines.

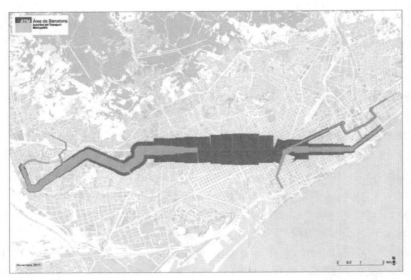

Source: Metropolitan Transport Authority, area of Barcelona.

Figure 7. Current Trambaix and Trambesós passenger numbers and forecast passengers for the Avenida Diagonal connection.

In large cities, the tram can also ensure a more efficient public transport network in city centres, as in Berlin, Sydney and Paris. The strategy for the T-3 line in Paris was to develop a circular line parallel to the Periferique using the interior boulevards. The aim was to offer greater passenger capacity in high density areas where the bus service was not competitive enough to lead people to stop using private vehicles.

Each city projects its tram lines according to the overground public transport system, replacing buses where they can no longer effectively deal with high demand. The case of Avenida Diagonal between Plaza Francesc Macià and Calle Valencia is a textbook case. Due to high demand that cannot be met under acceptable conditions, buses currently run at speeds below 9 km/h, with repeated bus-bunching episodes. The replacement of bus lines by a tramway within the framework of a perspective of the newly configured NXBus network compatible with the tram system is based on the criterion of maximum intermodality, just as in Amsterdam and Lyon, where the complementarity between different transport modes is ensured.

Trams potentially replace inefficient bus lines, transport many more passengers at a lower unit cost and free up resources for integrated systems that can be used to improve frequencies for key bus lines. The contribution of the tram to the efficiency of an integrated system is that it transports more and better at a lower cost.

The Tram as an Urban Planning Instrument

The reimplantation of the tram in major cities has led to urban planning on a large scale, both in terms of regenerating suburbs and refurbishing streets, central squares and buildings. In these projects, which typically go beyond merely refitting roads as tramways, new design criteria have been applied to improving the quality and quantity of the contiguous urban spaces. As for historic cities like Florence, the insertion of tramlines running through the old monumental quarter has had to be done in compliance with stringent criteria.

The image of city centres has changed greatly as a result of the entry into service of new tram systems. Major avenues, like Paseo de la Independencia and El Coso in Zaragoza, and squares, like Pey Berland and Comédie in Bordeaux, are outstanding examples of transformations resulting from tramway installation, as radical as they are successful. Perhaps the most spectacular image is that of Sydney, whose famous postcard of the 1980s featured the monorail against a skyline of skyscrapers. The monorail is no more and Sydney's centre now features tree-lined streets, restored brick buildings and open spaces reserved for people and trams. Sydney, indeed, exemplifies the new central urban space of the 21st century.

Avenida Diagonal, 11 km in length, is the longest street in Barcelona. It was a key element in Ildefonso Cerdà's 1859 proposal for development of the area now known as the Eixample. However, the Diagonal is not uniform along its length, as central sections and physiognomy are varied.

The 4-km long section between Plaza Francesc Macià and Plaza Glòries Catalanes is not homogeneous either, although its common section measures a uniform 50 metres between the building facades on either side. To its junction with Calle Sardenya, the avenue is structured in a regular and harmonious manner, with a symmetry emphasized by the monuments of Plaza Cinc d'Oros and Plaza Jacint Verdaguer. Lined with four rows of trees, it has a large central road with four traffic lanes and, on each side, a pedestrian boulevard, a side road and a narrow pavement next to the buildings. The avenue is featured by superb buildings that include Modernist houses like Casa de les Punxes, Noucentisme buildings like Ramon Llull School and other emblematic buildings, for instance that of the Banc de Sabadell. It is clear that any intervention that modifies this avenue and its elements needs to be planned and implemented with extreme care. In 2015, the section between Plaza Francesc Macià and Plaza Cinc d'Oros was partially reformed, to wide popular acclaim, given that the defining elements, trees and symmetry of this classical avenue were preserved while quality urban space was recovered for pedestrian use. However, it was extraordinarily timid in terms of tackling the supremacy of motorized transport: the central roadway remained exactly as it was.

The tram connection project proposes a more radical transformation in terms of recovering space for sustainable mobility by reserving the entire central roadway for this purpose. Taking full advantage of the implementation of a new infrastructure, the idea is to develop this section so that it is more in line with the new urban model envisaged for the Barcelona of the near future (see Figure 8). The criteria fulfil requirements regarding maintenance of the exceptional quality of Avenida Diagonal as an urban environment. No catenary will be used for this section (like in the central sections of the Bordeaux, Nice, Rennes, Zaragoza and Seville tramways) and a new symmetry is proposed for roundabouts, based on locating tramlines on the side of the sea and bike paths on the side of the mountain.

The connected-up tram system will be the great urbanizing catalyst for Barcelona — the pretext for transforming currently highly traffic-congested urban spaces into more human spaces. This prioritization of criteria of sustainable mobility will entirely undo the dimensions prescribed by 20th century traffic engineering manuals.

Source: Informative study on connecting the tram networks. TPO-SENER-TYPSA.

Figure 8. Image of the new section of Avenida Diagonal.

The Tram as the Engine for Environmental Policies

Many cities throughout the world are implementing sustainable mobility policies based on modernizing and extending tram networks. Berlin is a reference, given the ambition of its approach: a plan has been drawn up to build new lines, unify and modernize existing networks in the Eastern and Western sectors, replace rolling stock and extend the system along the main avenues of West Berlin. In a city which already has an excellent collective transport system, the tram project reflects a clear commitment to environmental principles. A message is being transmitted that urban space will be taken back from the car and that car use (especially diesel and heat engine vehicles) in the city will be gradually restricted, in the interest of both reducing local air pollution and contributing to the global fight against climate change. Data for Zaragoza pointing to a notable decrease in emissions and a great improvement in air quality confirm the environmental efficiency of the tram.

For cities like Barcelona with serious levels of urban pollution, mobility plans have already underlined the urgent need to cut back on the number of vehicles entering the city. The current plan aims for a reduction of 21%, which will require simultaneous action on several fronts. One strategy is to ensure that overground public transport is capable of competing with private cars in terms of replacing them or rationalizing their use as a mode of transport. Two important projects respond to this objective: completion of the NXBus overhaul of the bus system scheduled for 2018 and linkage between the two existing tram networks. Both projects aim to ensure an optimally efficient public transport system. The tramway connection and the Avenida Diagonal transformation project will prove to be of enormous pedagogical value in demonstrating that changing how urban space is managed and used can promote a culture of sustainability in a city.

Implementation of a tram system offers many benefits from an environmental point of view. Its contribution to reducing emissions is quantifiable and invariably significant, but its greatest contribution to sustainability lies in a new culture of mobility and the reimagining of urban space in terms of new priorities for its use.

Conclusion[13]

The return of the tram, often controversial and heatedly debated, is a reality in several cities and has brought about numerous positive changes. The cities — diverse in terms of size, morphologies and layout — include those which, with foresight, maintained their networks of the past and

---

[13] In this section Antonio González, Pere Macias, Abel Ortego, Elisabet Roca and Míriam Villares have contributed.

those that have taken advantage of the construction of a new tramway to implement urban renewal projects and improve connectivity between centres and suburbs.

The experiences of these cities point to a high-capacity transport mode that is efficient, easily accessible, affordable and comfortable. The urban planning potential of tramways is unquestionable, reflected as it is in improvements to the urban landscape, recovered public space, more sustainable mobility, densification and the construction of new amenities, not to mention the enhanced connectivity with outlying areas. As well as enhancing mobility, this kind of urban transformation based on tram systems reduces greenhouse gas emissions and acoustic pollution. In improving the quality of urban life, it is a key element in the environmental recovery of cities.

The case studies of Bordeaux, Lyon, Zaragoza, Sydney, Rotterdam, Amsterdam and Florence illustrate tram system challenges, solutions and enhancements that will contribute knowledge and experience to the Avenue Diagonal tram connection project. As can be seen from those cities, the implementation of a new tram system is a major opportunity for developing sustainable urban mobility policies and recovering public space from vehicles.

Barcelona is a city that already has consolidated underground and suburban railway systems. The tram network, however, is a key additional element in its systemic vision of public transport as a powerful means for achieving sustainable mobility. The connection of the two existing tram lines via the central section of Avenida Diagonal will, in conjunction with the newly configured NXBus network, enhance the rationality and efficiency of overground transport by more efficiently meeting mobility demands in a highly congested corridor.

This systemic vision also applies to mobility as a whole — with all its consequences — in which the tram will have a particular role to play both in the city centre and the entire metropolitan area. In other words, it is not a question of articulating the entire transport system around the tram, but of recognizing trams as the best solution for displacements that converge in a highly congested centre, while simultaneously responding to the need to recover the quality of the urban space along and in the vicinity of Avenida Diagonal. The potential of light-rail or tram systems, in terms of transport capacity and efficiency, urban renewal and environmental improvements, are evident in the Barcelona project, designed to radically transform its main artery by improving mobility and reducing contamination levels. However, its main value lies in the high pedagogical value of the action, which fits within the framework of a planning policy focused on urban recovery and sustainable mobility that also contemplates additional initiatives, such as the creation of "superilles" (traffic-calmed and pedestrian-only areas), the recovery of public space and the construction of an extensive network of bike lanes. The overall vision is to change the functional specialization of streets so as to rationalize the use of space, restrict the space given over to cars and prioritize the reclamation of public space — in quantitative and qualitative terms — for pedestrians, cyclists and public transport.

The vision, radical, transcendent and highly significant, is to switch from a situation of permanent congestion by some 60,000 vehicles (72,000 people) to a situation of more efficient and sustainable public transport that will guarantee greater capacity, shorter travel times and a significant reduction in emissions. The tram is not only the instrument of this transformation, but also its best image.

In short, Barcelona's overground intermodal transport system, articulated around comfortable and efficient trams, is a highly attractive public transport option. In light of the experiences of cities that have implemented similar systems, there is little doubt as to the capacity of a tram system to transform a city, enhance its landscape, improve its environment, pacify congested areas and recover public space for healthy, tranquil enjoyment by a city's inhabitants.

REFERENCES

Carmona, G. (2015). *Tranvias y otros despilfarros*. España: Madrid [*Trams and other economic wastes.* Spain: Madrid].

CERTU. (2002). *Evaluation des transports en commun en site propre, recommandations pour l'évaluation socio-économique des projets.* [*Evaluation of public transport in own site, recommendations for socio-economic evaluation of projects*].

Lois Gonzalez, R. C., Pazos Oton, M., Wolff, J. O. (2013). "Le tramway entre politique de transport et réhabilitation urbanistique dans quelques pays européens: Allemagne, Espagne, France, Suisse" In *Annales de géographie* 694: 619-643. ["The tramway between transport policy and urban regeneration in some European countries: Germany, Spain, France, Swiss" In *Annals of Geography* 694: 619-643].

Miglietti, A. (2008). *The role of participation in the implementation of sustainable mass transport measures. The case of Florence tramway system.* Graduation Thesis, Maastrich Graduate School of Governance.

Moreno, J. (2014). *Esquinas territoriales. Movilidad y planificación territorial, un modelo de integración: el Randstad-Holland.* Tesis doctoral, Universitat Politècnica de Catalunya. [*Territorial corners. Mobility and territorial planning, an integration model: Randstad-Holland.* Graduation Thesis, Universitat Politècnica de Catalunya].

Munro-Clark, M., and Thorne, R. (1989). "Hallmark events as an excuse for autocracy in urban planning: a case history of Sydney's Monorail." In *The Planning and Evaluation of Hallmark Events* edited by Syme, G. T., Shaw, B. J., Fenton, D. M. and Mueller, W. S., 154-171.

Offner, J. M. (2001). "Raisons politiques et grands projets." *Annales des ponts et chaussées* 99:55-59. ["Political reasons and big projects." *Annals of bridges and causeways* 99:55-59].

Ortego, A., Valero, A., Abadías, A. (2017). "Environmental Impacts of Promoting New Public Transport Systems in Urban Mobility: A Case Study." *Journal of Sustainable Development of Energy, Water and Environment Systems* 5:377-395. doi: http://dx.doi.org/10.13044/j.sdewes.d5.0143.

Proyecto Z2020xMUS, (2014), Financiado por la Agencia de Medioambiente y sostenibilidad del Ayuntamiento de Zaragoza. [Funded by the Environmental and Sustainability Agency of the City of Zaragoza]

Salmerón, J.C. (2017). *Tranvies 2017. La mobilitat urbana del segle 21.* Barcelona: Terminus. [*Tramways 2017. Urban mobility of s. 21.* Barcelona: Terminus].

Transport of NSW. *Sydney Lighht Rail* http://sydneylightrail.transport.nsw.gov.au/

About the Editor

Saúl Antonio Obregón-Biosca obtained the PhD and Master degree in the Polytechnic University of Catalonia, civil engineer in the Autonomous University of Querétaro (UAQ). Is member of the Mexican National System of Researchers, level 2; obtain the Mexican National award "José Carreño Romaní 2016" by the Ministry of transport and communications of Mexico and the Mexican Association of Transport Engineering; Member of the American Society of Civil Engineers; ID: 9306847. Since 2009 has been working as professor-researcher in the Graduate Studies Division, Engineering Faculty in the UAQ and was Chief of the Master program in Transport Engineering. His research activity has been focused on the engineering of the infrastructure networks in the territory in two areas: first, the transport infrastructure as the backbone of the territory, the second metropolitan mobility.

INDEX

A

access, 3, 5, 6, 12, 17, 18, 21, 23, 24, 29, 36, 109, 139, 170
accessibility, v, vii, viii, 1, 2, 3, 4, 5, 6, 7, 8, 9, 10, 13, 14, 15, 18, 19, 20, 21, 28, 29, 30, 31, 32, 33, 34, 35, 36, 38, 39, 40, 41, 43, 153, 154, 167
accessibility index, 5, 6, 7
active mobility, viii, 45, 46, 53, 56, 57, 58, 60, 61, 62, 63, 64, 65, 66, 67, 68
agencies, 29, 63, 73, 79, 89, 104, 111, 118
air quality, 156, 166, 173, 178
airport, v, 27, 34, 103, 104, 106, 107, 108, 109, 110, 111, 113, 114, 115, 116, 118, 119, 120, 160
alternative energy, 52
architectural space, 54
assembly of cars, 53
assessment, v, vi, viii, 4, 10, 36, 71, 72, 85, 86, 89, 90, 91, 94, 95, 96, 100, 110, 121, 122, 123, 125, 126, 127, 128, 134, 136
asset, 25, 72, 73, 74, 75, 76, 85, 86, 87, 88, 89, 99, 100, 101, 127, 144
availability, 7, 29, 47, 65, 145

B

bike lanes, 62, 63, 155, 180
bikeways, 62, 65
budget allocation, 72, 81
business model, 52
businesses, 19, 40, 62, 161, 165

C

case studies, viii, 20, 171, 179
centrifugal forces, 17
centripetal forces, 17
challenges, viii, 2, 72, 86, 103, 118, 122, 127, 145, 146, 159, 179
cities, vii, 2, 11, 13, 16, 22, 24, 32, 46, 47, 49, 50, 51, 53, 55, 57, 58, 59, 60, 61, 62, 63, 64, 65, 66, 68, 69, 150, 152, 153, 155, 157, 158, 159, 161, 162, 171, 173, 174, 175, 177, 178, 179, 180
citizens, 3, 12, 47, 62, 63, 64, 66, 155, 158, 161, 170, 171
climate change, 157, 178
collective mobility, 48, 56, 63
collective transport, 53, 55, 64, 161, 177

commercial, 2, 10, 22, 25, 118, 152, 156, 158, 160
communication, 63, 64, 66, 75, 87, 88, 92
communication skills, 87
communication strategies, 75
comparative advantage, 137
complexity, 10, 12, 15, 67, 75, 79, 80
computer, 103, 104, 107, 131
configuration, 113, 127, 139, 141, 142, 171
connectivity, 8, 9, 146, 156, 159, 167, 179
construction, 1, 19, 42, 62, 63, 64, 66, 77, 90, 92, 105, 137, 153, 160, 168, 170, 179, 180
consumers, 6, 23, 24, 47
consumption, 16, 23, 26, 157
consumption patterns, 23
cost, 3, 4, 5, 6, 8, 9, 12, 14, 17, 18, 21, 25, 26, 27, 49, 64, 73, 78, 88, 89, 114, 116, 121, 123, 125, 143, 147, 151, 152, 153, 156, 168, 172, 175
Costa Rica, v, 71, 72, 73, 77, 78, 89, 90, 93, 94, 96, 97, 98, 99, 100, 101
cost-benefit analysis, 147
crisis management, 139
critical infrastructure, 141
cultural factors, 52
cultural trends, 52
cycle paths, 59
cycling infrastructure, 64
cyclists, 15, 59, 62, 153, 156, 180

D

damages, 82, 122, 143, 145
data bases, 72
data collection, 72, 73, 79, 87
data processing, 80
database, 104, 107, 117
decision makers, 103, 117
decision-making process, 29, 72, 76, 95, 115

deficiencies, 82, 85, 143
deflectometer, 71, 82, 91
density, 18, 19, 22, 42, 47, 54, 164, 174
distress, 110, 113, 114

E

economic, vii, viii, 1, 2, 3, 12, 14, 17, 19, 20, 21, 23, 28, 29, 31, 32, 36, 38, 40, 42, 48, 49, 50, 53, 59, 61, 78, 88, 89, 95, 110, 114, 119, 122, 123, 126, 127, 136, 144, 146, 159, 172, 180, 181
economic activity, 17, 21, 29
economic consequences, 119
economic crisis, 172
economic development, 1, 19, 20, 21, 31, 40
economic evaluation, 181
economic growth, 2, 20
economic problem, 14
economic relations, 123
economic resources, 110, 114, 119
economic welfare, 28
economic well-being, 3
economies of scale, 17
economy, 22, 32, 35, 38, 39, 41, 47, 65, 121, 125, 127, 172
employment, vii, 8, 9, 12, 20, 29, 31, 40, 42
employment growth, vii, 10, 12, 40
energy, 22, 39, 139, 144, 154, 156, 165, 166
energy consumption, 22, 39, 139, 157, 166
engineering, 73, 80, 89, 100, 140, 177
environment, 3, 4, 9, 13, 24, 27, 30, 32, 51, 54, 55, 57, 58, 60, 62, 63, 66, 124, 127, 166, 171, 176, 180
environmental change, 20
environmental conditions, 76
environmental effects, 124
environmental factors, 25, 28
environmental impact, 121, 144, 168
environmental issues, 51
environmental perception, 57

Index 187

environmental psychology, 54
environmental quality, 154, 155
environments, 14, 32, 54, 60, 65, 69, 151
evaluation, 7, 20, 41, 72, 81, 82, 83, 85, 90, 91, 92, 93, 94, 95, 100, 101, 104, 106, 108, 110, 111, 116, 117, 118, 119, 158, 181

F

financial cost for transport, 64
financial crisis, 151
freight transport, 25
friction, 5, 6, 80, 84, 111, 112, 113, 115, 118, 123, 128, 129, 133
fuel consumption, 14, 28
funds, 19, 72, 77, 114, 118

G

geographical, 6, 10, 11, 29, 30, 48, 49, 146, 158
geography, 1, 17, 23, 35, 36
governments, 12, 52, 65
greenhouse gas, 22, 154, 156, 179
greenhouse gas emissions, 22, 156, 179

H

habitability, 46, 51, 54, 58, 59, 62, 63, 65, 66, 171
habitability of cities, 61, 63, 66
hazardous materials, 135
hazardous substance, 136, 137, 146
highways, 19, 20, 25, 139, 167
history, 106, 107, 109, 110, 181
housing, vii, 2, 3, 15, 19, 23, 33, 48, 154
human, 8, 35, 46, 54, 55, 57, 58, 59, 60, 61, 64, 67, 74, 75, 115, 176
human activity, 59

human behavior, 46, 60
human interaction, 55
human relationships, 46
human resources, 74
hybrid vehicles, 52

I

idiosyncrasy, 45, 46, 47, 53
implementation, 72, 73, 75, 85, 86, 87, 88, 89, 99, 107, 118, 119, 120, 145, 156, 159, 160, 161, 170, 171, 172, 173, 176, 178, 179, 181
improvements, 9, 17, 31, 154, 179, 180
income, 4, 10, 15, 16, 18, 19, 21, 24, 33, 49, 50
individual idiosyncrasy, 51
individual perception, 5
individuals, 4, 5, 6, 7, 9, 10, 23, 37, 46, 47, 48, 49, 54, 55, 56, 62, 63, 66
industrial location, vii, 17, 18, 29
infrastructure, vi, vii, viii, 1, 2, 3, 9, 11, 13, 17, 18, 20, 21, 24, 25, 28, 29, 34, 37, 40, 46, 47, 49, 55, 59, 60, 61, 62, 63, 64, 65, 66, 72, 87, 88, 89, 92, 95, 112, 121, 122, 123, 124, 125, 126, 127, 128, 129, 132, 134, 136, 137, 139, 140, 141, 143, 144, 145, 146, 147, 152, 153, 154, 155, 176, 183
infrastructure in mobility, 49
inhabiting, 57, 58
inventory, 78, 79, 88, 106, 107, 108, 117, 118
investment, 3, 12, 19, 21, 28, 31, 48, 53, 65, 74, 77, 81, 96, 97, 98, 100, 101, 118, 123, 126, 160
IRI (international roughness index), 71, 83, 92, 93, 94, 95, 96, 100, 112

L

land use, viii, 1, 6, 11, 13, 18, 23, 24, 29, 34, 38, 41, 43, 67
local market, 18, 53

M

maintenance, 25, 72, 73, 77, 78, 81, 90, 94, 95, 96, 103, 104, 105, 106, 107, 108, 109, 110, 112, 113, 114, 115, 116, 117, 118, 119, 120, 121, 122, 123, 124, 125, 126, 127, 128, 129, 131, 134, 135, 136, 140, 141, 142, 143, 144, 145, 146, 147, 168, 176
management, v, vi, viii, 1, 25, 32, 41, 43, 71, 72, 73, 74, 75, 76, 77, 78, 79, 80, 81, 85, 86, 87, 88, 89, 90, 92, 94, 95, 97, 98, 99, 100, 101, 103, 105, 106, 107, 108, 110, 114, 116, 118, 119, 120, 121, 122, 123, 125, 127, 134, 139, 143, 159, 161, 168
manufacturing, 9, 18, 53, 127
marginalized neighborhoods, 50
market access, 17
market structure, 28
means of transport, 22, 46, 47, 58, 62, 65, 156, 158
metropolitan areas, 11, 14, 16, 23, 173
mobility, v, vi, vii, 1, 3, 4, 7, 8, 9, 12, 13, 14, 15, 16, 21, 22, 29, 31, 35, 36, 37, 39, 40, 42, 43, 45, 46, 47, 48, 49, 50, 51, 52, 53, 55, 56, 57, 58, 59, 60, 61, 62, 65, 66, 67, 68, 69, 77, 101, 149, 150, 153, 155, 156, 157, 158, 159, 161, 162, 163, 164, 167, 170, 172, 173, 176, 177, 178, 179, 181, 182, 183
monocentric model, 8, 10
motor vehicles, 52
motorized mobility, 61, 65
motorized vehicle, 59, 62, 65

N

neighborhood, 19, 36, 47, 51
neighborhood characteristics, 19
network level, 72, 77, 79, 80, 95, 105, 106
new economic geography, 1, 17
non-motorized transport, 66

O

operations, 28, 76, 111, 112, 122, 123, 124, 125, 127, 128, 129, 136, 138, 140, 141, 142, 143, 169
opportunities, 2, 5, 6, 9, 11, 16, 19, 33, 36, 59, 60, 61
oversight, 73, 89, 90, 95, 96, 97, 99, 100

P

pavement, v, viii, 25, 57, 71, 72, 76, 77, 78, 79, 80, 81, 82, 84, 85, 86, 87, 88, 89, 90, 91, 92, 93, 94, 95, 96, 97, 98, 99, 101, 103, 104, 105, 106, 107, 108, 109, 110, 111, 112, 113, 114, 116, 117, 118, 119, 120, 130, 153, 176
pavement management system, 71, 72, 76, 77, 78, 79, 80, 85, 86, 87, 89, 98, 99, 103, 104, 106, 116, 118, 120
pedestrian traffic, 62, 64
pedestrians, 15, 59, 62, 153, 155, 156, 161, 180
peripheral neighborhoods, 49
peri-urban, 12, 16, 41
physical environment, 4, 25, 59
planning, 1, 2, 4, 9, 10, 11, 12, 22, 25, 29, 31, 33, 34, 36, 38, 39, 40, 41, 42, 43, 62, 63, 74, 75, 79, 80, 81, 86, 90, 92, 118, 144, 149, 150, 151, 153, 158, 161, 162, 163, 168, 175, 179, 180, 181
plants, 53, 153, 165, 167

Index

policies, vi, 9, 11, 12, 16, 22, 25, 34, 38, 41, 50, 52, 72, 73, 74, 104, 106, 107, 113, 114, 117, 123, 149, 173, 177, 179
policy, 4, 11, 21, 25, 39, 113, 115, 180, 181
polycentric model, 8, 10, 34
population, vii, 2, 3, 6, 7, 10, 11, 12, 15, 16, 20, 21, 22, 29, 48, 52, 57, 61, 162, 169
private, 12, 22, 25, 48, 49, 50, 52, 55, 56, 59, 61, 62, 65, 66, 73, 144, 152, 154, 156, 158, 160, 165, 167, 168, 169, 170, 174, 178
private cars, 48, 50, 160, 178
private levels, 52
private mobility, 50, 56
private transport, 48, 49, 50, 52, 56
private transportation, 56
private vehicle, 22, 25, 48, 62, 65, 66, 152, 156, 158, 160, 169, 170, 174
project, 31, 72, 73, 74, 75, 77, 78, 79, 80, 83, 85, 86, 88, 105, 135, 139, 140, 145, 150, 151, 155, 161, 163, 164, 170, 172, 176, 177, 178, 179, 180
psychological activities, 54
public and private interests, 53
public and private transport, 49
public institutions, 52
public means of transportation, 49
public open spaces, 57
public plaza, 54
public policies, 12, 22, 49, 62, 63, 65
public space, 49, 50, 51, 55, 56, 59, 62, 63, 66, 150, 151, 152, 153, 155, 161, 165, 173, 179, 180
public transport, vii, 1, 4, 12, 14, 15, 16, 21, 36, 50, 52, 53, 56, 64, 66, 150, 151, 152, 154, 156, 158, 161, 164, 170, 172, 173, 174, 178, 179, 180, 181
public works, 63, 65, 68

R

rail operators, 27
railroads, 32, 53
railway transportation, 122, 127, 133, 135, 136, 144
related activities, 47
resistance, 71, 81, 84, 86, 93, 95, 99, 100, 137
resource utilization, 13
resources, 8, 73, 87, 88, 92, 97, 126, 127, 144, 175
right to the city, 63
road network, v, vii, 10, 20, 25, 29, 71, 72, 73, 77, 78, 79, 81, 83, 85, 87, 88, 89, 90, 91, 92, 93, 94, 95, 96, 97, 98, 99, 100, 101, 163
roads, viii, 20, 28, 34, 39, 46, 49, 52, 55, 56, 60, 62, 63, 64, 74, 75, 77, 90, 91, 93, 95, 108, 110, 119, 120, 139, 140, 146, 167, 173, 175

S

safety, 25, 51, 76, 77, 84, 85, 93, 111, 112, 121, 122, 123, 125, 126, 135, 137, 140, 169
sense of belonging, 52, 55, 57, 60, 61, 66
sense of community, 46, 56, 57, 58, 60
shared spaces, 55
shipping industry, 28
sidewalks, 59, 60, 62, 63, 65
social bond, 47
social dimension, 2, 54
social interaction, 32, 51, 57, 64
social mobility, 48
socioeconomic effects, 2
socio-economic layers, 48
substantially, 57
subway systems, 53

sustainability, vi, 22, 68, 75, 121, 127, 134, 136, 144, 149, 150, 178, 181

T

territorial, viii, 1, 2, 4, 8, 14, 15, 16, 20, 22, 28, 29, 37, 161, 168, 181
territorial transformation, viii, 1, 2, 8, 20
territory, vii, 2, 8, 11, 20, 21, 23, 28, 29, 30, 154, 163, 167
theoretical approach, 145
theoretical approaches, 145
tramway, 150, 151, 154, 155, 165, 168, 170, 174, 175, 178, 179, 181
transformation, viii, 1, 2, 20, 67, 139, 150, 176, 178, 179, 180
transmilenio, 64
transport infrastructure, vii, viii, 1, 2, 3, 15, 17, 18, 20, 21, 24, 28, 29, 31, 32, 37, 183
transport modelling, 24, 29, 31
transport system, vii, 3, 4, 7, 8, 9, 10, 19, 20, 21, 50, 122, 125, 145, 155, 158, 159, 163, 172, 173, 178, 179, 180
transportation asset management, 71, 72, 73, 74, 76, 86, 88, 98, 100
transportation means, 51
transportation systems, 10, 150
travel cost, 3, 6, 9, 21, 29
travel demand, vii, 22, 24, 25, 38

U

unequal societies, 48
urban, v, vi, viii, 1, 2, 3, 8, 9, 10, 11, 13, 14, 15, 16, 19, 21, 22, 23, 24, 28, 29, 31, 32, 33, 34, 35, 36, 37, 38, 39, 40, 41, 42, 43, 45, 46, 47, 53, 54, 55, 57, 59, 60, 61, 62, 63, 64, 65, 66, 67, 68, 69, 149, 150, 151, 152, 153, 154, 156, 158, 161, 162, 163, 164, 165, 169, 171, 173, 175, 176, 177, 178, 179, 180, 181, 182
urban community, 46
urban design, 16, 59, 62, 63, 64, 161
urban designers, 59, 62
urban development plans, 65
urban dimension, 54
urban environment, 13, 55, 57, 60, 62, 63, 176
urban growth, 2, 31, 32, 33, 43, 66
urban habitability, 46, 54, 60, 61
urban infrastructure, 60, 61, 62, 63, 64
urban networks, 59, 62
urban planners, 59, 65
urban policies, viii, 65
urban project, 150, 152, 153, 161, 171
urban realm, 45, 54
urban regulation, 63
urban sprawl, 8, 11, 13, 14, 16, 29, 31, 32, 33, 36, 37, 41, 43, 63
urban sprawl growth, 63
urban surplus value, 62
utility theory, 7

V

value of time, 50
values, 1, 2, 29, 41, 59, 93, 95, 117
vehicles, 14, 22, 25, 27, 48, 51, 52, 53, 59, 62, 64, 65, 66, 82, 83, 85, 91, 137, 139, 140, 156, 158, 160, 169, 170, 174, 178, 179, 180
vibration, 122, 133
vision, 12, 57, 58, 59, 161, 179, 180

W

walkways, 62

Road Traffic and Safety

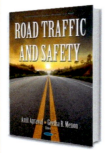

EDITORS: Amit Agrawal and Geetha R. Menon (Department of Neurosurgery, Narayana Medical Hospital and College, Pradesh, India)

SERIES: Transportation Issues, Policies and R&D

BOOK DESCRIPTION: The present book consists of fifteen chapters related to various aspects concerning road traffic and safety, including epidemiology of road traffic injuries, occupant protection and safety devices, risk factors, a manual of safety measures, road safety in hilly terrain and conflict zones, prevention of head injuries, the role of alcohol and bicycle related injuries.

HARDCOVER ISBN: 978-1-53612-489-7
RETAIL PRICE: $195

U.S. Transit, Transportation and Infrastructure: Considerations and Developments. Volume 7

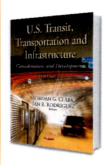

EDITORS: Jordan G. Clark and Ian R. Rodriguez

SERIES: U.S. Transit, Transportation and Infrastructure: Considerations and Developments

BOOK DESCRIPTION: Policymakers at all levels of government are debating a wide range of options for addressing the nation's faltering economic conditions. This book examines policy issues associated with using infrastructure as a mechanism to benefit economic recovery.

HARDCOVER ISBN: 978-1-63485-447-4
RETAIL PRICE: $150

Road User Charges Based on Mileage: Considerations and Viability

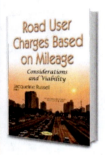

Editor: Jacqueline Russell

Series: Transportation Issues, Policies and R&D

Book Description: A mileage-based road user charge would involve assessing owners of individual vehicles on a per-mile basis for the distance the vehicle is driven. This book examines considerations and viability of road user charges based on mileage.

Softcover ISBN: 978-1-53610-498-1
Retail Price: $62

Making Federal Highway Spending More Productive: Analyses, Approaches and Perspectives

Editor: Rickey Lambert

Series: Transportation Issues, Policies and R&D

Book Description: Federal spending on highways totaled $46 billion in 2014, roughly a quarter of total public spending on highways. This book discusses approaches to making federal highway spending more productive, as well as the status of the Highway Trust Fund and options for paying for highway spending.

Hardcover ISBN: 978-1-53610-314-4
Retail Price: $125

THE LAW OF AIR, ROAD AND SEA TRANSPORTATION

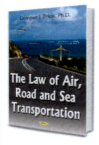

AUTHOR: Georgios I. Zekos, Ph.D. (Advocate and Economist, TEI of Central Macedonia, Serres-Macedonia-Hellas, Greece)

SERIES: Transportation Issues, Policies and R&D

BOOK DESCRIPTION: International markets expand through better communication and transport technology. Transport networks are at the heart of the supply chain and are the foundation of any country's economy. The author discusses these issues and more in this important book on international markets and transportation.

HARDCOVER ISBN: 978-1-63485-740-6
RETAIL PRICE: $230

TRAFFIC ACCIDENTS AND SAFETY: NEW RESEARCH

EDITOR: Garrett Bowman

SERIES: Transportation Issues, Policies and R&D

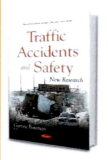

BOOK DESCRIPTION: Traffic accidents (TAs) represent a significant public health issue and are associated with behavioral factors, vehicles safeties and conditions of the urban spaces. According to the Centers for Disease Control and Prevention, road crashes are one of the major causes of morbidity and mortality in the United States. This book provides new research on traffic accidents and safety.

SOFTCOVER ISBN: 978-1-63485-517-4
RETAIL PRICE: $82